T0181619

Construction Auditing

Peter Wotschke · Gregor Kindermann

Construction Auditing

Planning - Implementation - Use

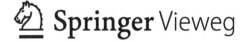 Springer Vieweg

Peter Wotschke
Hochschule für Wirtschaft und Recht
Berlin, Germany

Gregor Kindermann
BMC – Baumanagement & Controlling AG
Berlin, Germany

ISBN 978-3-658-38840-9   ISBN 978-3-658-38841-6   (eBook)
https://doi.org/10.1007/978-3-658-38841-6

This Springer Vieweg imprint is published by the registered company Springer Fachmedien Wiesbaden GmbH, part of Springer Nature.
The registered company address is: Abraham-Lincoln-Str. 46, 65189 Wiesbaden, Germany

Preface

In one of the first lecutures, our students at the Berlin School of Economics and Law (HWR) are asked about their motivation for having taken up the study of civil engineering. It is like a ritual when they are asked, "Why are you actually here?"

The initial laughter is usually followed by a few moments of reflection. When specifically addressed, one or the other declares that they want to build bridges, erect proud structures or take over their father's construction company. No one has yet expressed a desire to work in construction auditing.

Construction auditing does not seem to have a good lobby. As a subfield of internal auditing, this would not be surprising. After all, the auditor, as already humorously and impressively portrayed in the comedy of the same name by Nicolai W. Gogol from 1836, is often seen as a spectre and threat.

Perhaps it leads a shadowy existence in the large number of possible fields of activity of civil engineers in traffic route construction, civil engineering, structural engineering, hydraulic engineering or construction operations. Civil engineers work as specialist planners, as construction or project managers, as managers or consultants – but as auditors?

With this book, we would like to make a contribution to changing this image and provide information about the exciting, diverse and varied field of activity in construction auditing.

The term audit is derived from Latin: re- (again, back) and videre (to look at), i.e. to look at again, to look back, to review. This review can include technical, commercial or legal aspects and thus goes beyond a business review of company processes on file or the uncovering of misconduct relevant under criminal law.

This book is the result of many years of consulting work in the areas of construction management and controlling, especially in supplementary management, the investigation of disturbed construction processes and construction auditing. Our special thanks go first and foremost to our employees Manja Sondermann and Sandra Ziegler, who actively supported us in this consulting work and contributed to the success of this book.

And finally, we would like to thank our families, especially our wives and children, who had to spare us many an evening during the writing of the book and who nevertheless supported and encouraged us morally – and to some extent also in terms of content – to bring this project to a successful conclusion.

We hope you enjoy studying this book and that you gain the maximum amount of knowledge. We wish you much success in the implementation and look forward to your feedback.

Berlin, Germany   Peter Wotschke
January 2021   Gregor Kindermann

Contents

List of Figures

List of Tables

Basics

1

1.1 Quality Management

Quality means the totality of the characteristic properties and condition of an object or person.[1] There is no uniform definition of the term quality. Quality can be used both in a positive sense – something is of high quality – and in a negative sense – something is of poor or inferior quality.

With DIN EN ISO 9001, the term was defined internationally. According to this, quality is the "degree to which a set of inherent characteristics of an object fulfils requirements". In this context, "inherent" means "contained" or "included".[2]

Quality is something relative and describes the conformity of an object or person with given or self-determined requirements (cf. Fig. 1.1). Thus, quality can only be measured if the individual requirements are known or have been determined. Depending on the requirements of the user, the result can sometimes be assessed as "good" and sometimes as "bad", or all values in between.[3] The meaning is thus always dependent on the respective requirements of the user and the context from which the term quality is used. Depending on the user's requirements for quality, the term is associated with different contents and evaluations.[4]

The requirements are set by a wide variety of parties. DIN EN ISO 9001 speaks of "interested parties". The most obvious "party" is the customer who purchases a product or

---

[1] Duden Editorial (2020).

[2] DIN EN ISO 90001, p. 18.

[3] Brüggemann and Bremer (2020, pp. 3–4).

[4] Jakoby (2019, pp. 7–9).

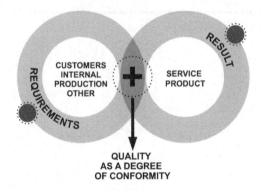

Fig. 1.1 Influences on the concept of quality

service. But also the different departments in the company, employees, suppliers or other involved persons have different demands on the quality.[5]

Most companies are successful on the global market when they manage to meet their customers' requirements for the quality of their products and services.[6] This success does not come about by chance, but is the result of a targeted planning and control process.

These processes are systematically managed by the quality management of a company.[7] One of the primary objectives of a company is therefore to define and meet the various requirements of the product or service.[8]

First of all, the customer's requirements for a product or service must be met as a matter of priority. However, other participants in the production process can also make demands on the quality of a product or service. This applies to all departments within the company, such as development, purchasing, production, but also to those involved outside the company, such as suppliers.[9]

Today, the idea of quality significantly determines the actions and goals of companies. The reasons for this are the ever-increasing global competition, rising customer expectations, the digital networking of customers worldwide as well as a higher complexity and variety of products and services.[10]

Thus, until the beginning of the twentieth century, the products were only manufactured within a manufactory or a craft business, often by only one or a few people. These

---

[5] Jakoby (2019, p. 10).

[6] Brüggemann and Bremer (2020, p. 1).

[7] Brüggemann and Bremer (2020, p. 124).

[8] Jakoby (2019, p. V).

[9] Jakoby (2019, p. 10).

[10] Brüggemann and Bremer (2020, pp. 1–3).

were then solely responsible for the quality. As industrialisation progressed and demand increased, the production strategy changed. The manufacturing process was taken over by specialized workers within the company or completely outsourced. This also changed the requirements for quality inspection.[11]

With such a specialised production split into individual production steps, the responsibility for a product shifts away from a single craftsman to many different people involved.[12]

In the beginning, it was sufficient to check the finished product for defects. If the requirements placed on the product were not met, the product was sorted out or reworked.

The higher the number of units, the more difficult and costly the final inspection of each individual product becomes. It therefore became necessary to review and improve the manufacturing processes from development and planning to delivery of the components and production. Each individual process step is optimized in such a way that a consistent quality of the final product can be ensured.[13] Quality inspection became quality management.

Quality management builds on the quality objectives and the quality policy of a company. It uses basic planning and control methods to align the company with these common quality objectives. Quality management plans and controls all processes that interact in the creation of a product or service. These can thus be continuously optimized and improved.[14]

Entire quality management systems (QMS) are created through the interaction of various methods that serve the planning and control of quality assurance processes.[15] The focus is on the functions of defining quality policy and quality objectives, quality planning, quality control, quality assurance and quality improvement (cf. Fig. 1.2), as they are also defined in the DIN EN ISO 9000 series of standards.[16]

The verification of the quality management systems can be checked in accordance with the DIN EN ISO 9000 standard by means of an investigation – a so-called audit. This is a systematic and process-independent examination of the quality-related activities in accordance with the planned instructions and objectives to be achieved.[17] It is determined whether conditions or processes correspond to a standard or were carried out in accordance with the standard.

To achieve this, internal audits are conducted in the company at planned intervals to verify that the quality management system in place meets the company's quality policy and objectives and realizes and maintains the requirements thereof.[18] The internal audits in a company serve to:

---

[11] Brüggemann and Bremer (2020, p. 5).

[12] Jakoby (2019, p. 21).

[13] Jakoby (2019, p. 8).

[14] Jakoby (2019, p. 9).

[15] Jakoby (2019, p. 128).

[16] DIN EN ISO 90001, p. 17.

[17] Brüggemann and Bremer (2020, p. 137).

[18] DIN EN ISO 90001, p. 41.

Fig. 1.2 Tasks of quality management

- Risk minimization and asset protection,
- Process reliability through defined procedures and controls,
- Prevention of fraudulent acts (white-collar crime),
- Realization of considerable savings potentials,
- Rejection of excessive costs and settlements,
- Quality improvement for the process flow,
- Transfer of best practice into the company.

In external audits, the company is examined by an external independent person, e.g. an auditor or a certification body, with the aim of e.g. certifying a product or process.

1.1.1   Internal Audit

In a company, the management assumes the leadership functions consisting of planning, organising and monitoring (cf. Fig. 1.3).

Due to the rapid growth, dynamics and complexity in today's globalized corporate world, it is of particular importance for the management to control and monitor the company holistically. Globalization creates risks and opportunities for the company, which must be identified, analyzed, evaluated and corrected or used.

In large companies, the process-independent monitoring function can no longer be covered by the management alone. The management therefore sets up an internal audit department which, as an instrument of the management, is mainly directly subordinate to the management.[19]

This is also clearly illustrated by the role that Internal Audit plays in the Three Lines of Defense model. The three-lines-of-defense model (cf. Fig. 1.4) can be used to pursue a systematic approach to risks that may arise in a company.[20]

---

[19] Bünis and Gossens (2016, p. 17).
[20] Berwanger and Kullmann (2012, p. 86).

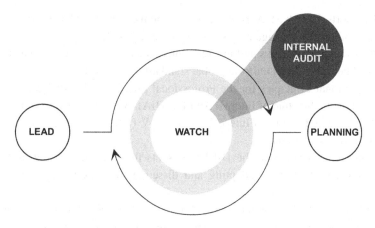

Fig. 1.3 Classification of internal auditing

Fig. 1.4 Three lines of defense model

The first line concerns the company's internal control system. Risks are identified, assessed and corrected here from operational day-to-day management.

In the second line, risk management systems support the activities of the first line and communicate them throughout the company.

The third line, in turn, monitors the company's risk management systems from the second line. This is the task of Internal Audit.[21] Internal Audit thus acts as an independent

---

[21] Schmidt (2016, p. 1).

body and supports the risk analysis of the first line at the operational level and reviews the effectiveness of the risk management systems of the second line.[22]

With the exception of banks and insurers, internal audit is not subject to any statutory regulations,[23] but can rely on uniform global standards in the performance of its duties.

The International Principles for the Professional Practice of Internal Auditing (IPPF) are published by the Institute of Internal Auditors (IIA). A translation into German is published by the German Institute of Internal Auditors e.V. (DIIR) – together with its Austrian and Swiss sister associations.

As a non-profit association, the DIIR represents the internal auditing profession in Germany and has the task of adapting and disseminating international standards for Germany.[24]

The IPPF makes the basic principles, definition, code of ethics and standards mandatory and describes the characteristics, procedures and activities of an internal audit function.[25]

It is imperative to ensure that the internal auditor is independent of day-to-day business, although he needs comprehensive insight into the company's processes for his auditing task.

The use of an internal audit helps the management to continuously improve the business processes through an independent and objective audit, to identify risks and to use opportunities[26] and thus to achieve the desired quality goals and quality assurance. Internal auditing creates added value for the company by monitoring and controlling internal processes and organisational structures with regard to their correctness, regularity, expediency and economic efficiency.[27]

In carrying out its activities, Internal Audit is guided by the quality requirements of its stakeholders or "interested parties", as they are referred to in the IIA/DIIR definition. In addition to the management, these are also other persons, companies or organisations involved in a company, such as employees, suppliers, owners and shareholders.[28]

This leads to the following definition of internal audit according to the IAA/DIIR:[29]

Internal Audit provides independent and objective assurance and consulting services designed to add value and improve business processes. It assists the organization in achieving its objectives by using a systematic and focused approach to evaluate and help improve the effectiveness of risk management, controls, and governance processes.

---

[22] Eulerich (2018, pp. 37–38).
[23] Otremba (2016, pp. 92–95).
[24] Berwanger and Kullmann (2012, p. 29).
[25] Bünis and Gossens (2016, p. 19).
[26] Eulerich (2018, pp. 11–12).
[27] DIIR MaRisk Working Group (2019, p. 5).
[28] DIIR MaRisk Working Group (2019, p. 6).
[29] DIIR MaRisk Working Group (2019, p. 5).

The IIA/DIIR definition thus clearly specifies the audit areas in which internal auditing should be active, the tasks and objectives of internal auditing, and the principles on which internal auditing should operate. These binding criteria are briefly described below.[30]

The auditing activities of Internal Audit are determined by quality management, from which the assurance objectives are derived. The assurance objectives are used to develop the audit objectives and audit assignments. The process areas to be audited are subsequently derived from these. The audit objectives are based on the overriding objectives of the company management within the framework of quality assurance and quality policy.[31]

The audit fields of the internal audit include typically

- Financial auditing,
- Operational auditing, and
- Managerial auditing.

In addition, new areas of auditing have opened up in more recent times due to the increased demands placed on companies as a result of globalisation. These include.[32]

- Compliance audits,
- Internal consulting,
- Risk management and
- the detection of fraudulent acts.

The fields are shown in Fig. 1.5 merged into a model and are briefly explained below.

One of the most classic audit fields of internal **auditing** is **financial** auditing. The internal audit department examines and evaluates the company's financial and accounting systems and the corresponding internal control system for appropriateness, accuracy, reliability and compliance. This mainly involves ex-post observations.[33]

After the financial audit, the **operational audit** is the second classic audit field. Here, the audit and evaluation of corporate structures and processes is carried out for expediency and cost-effectiveness with regard to corporate goals and efficiency.[34] Operational audits are ex-ante observations with the aim of examining, evaluating and improving corporate processes and structures. The focus is on the effectiveness and efficiency of all work processes and procedures in all areas of the company.[35]

---

[30] Eulerich (2018, p. 3).

[31] Bünis and Gossens (2016, p. 16).

[32] Berwanger and Kullmann (2012, pp. 78–79); Eulerich (2018, p. 108).

[33] Berwanger and Kullmann (2012, p. 42).

[34] Berwanger and Kullmann (2012, pp. 47–48).

[35] Berwanger and Kullmann (2012, pp. 81–82).

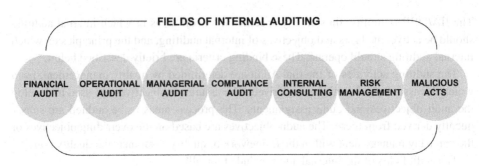

Fig. 1.5 Fields of internal auditing

The **managerial audit** serves to examine the appropriateness of management decisions. It answers the question of whether management is using and executing suitable planning, control and monitoring instruments in a targeted manner.[36]

Corporate compliance is the duty of corporate bodies and employees to comply with generally applicable laws and guidelines, but also with internal company rules and regulations. As part of the compliance audit, the internal audit department examines the company's processes with regard to compliance with statutory and internal company rules and regulations.[37]

Another task of Internal Audit is to perform an **advisory function**. As a result of its extensive auditing activities, Internal Audit has built up extensive knowledge of structures and processes within the company. Similarly, findings and recommendations regularly form part of the final audit report as part of the audits.[38]

As part of **risk management**, Internal Audit reviews the proper implementation of internal guidelines and the completeness and assessment of identified corporate risks as the second line of defense. The suitability and effectiveness of the risk management system are reviewed as the third line of the three lines of defence model described.

Due to the increase in **fraudulent acts** within the globalized business world, companies must also increasingly deal with this topic. Dolose acts include all acts within the scope of white-collar crime that are intentionally carried out to the detriment of the company.

Within the framework of new laws, companies must increasingly implement measures to detect and prevent these fraudulent acts. These tasks are often assumed by the internal audit department.[39] In this context, it supports the development and implementation of measures for prevention, detection and clarification.[40]

---

[36] Berwanger and Kullmann (2012, pp. 82–83).

[37] Berwanger and Kullmann (2012, pp. 39–40).

[38] Berwanger and Kullmann (2012, p. 84).

[39] Berwanger and Kullmann (2012, pp. 41–42).

[40] Berwanger and Kullmann (2012, p. 87).

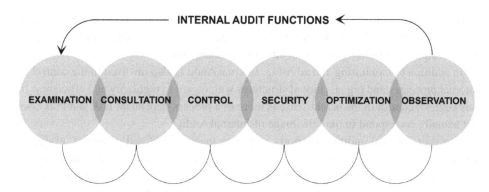

Fig. 1.6 Internal audit functions

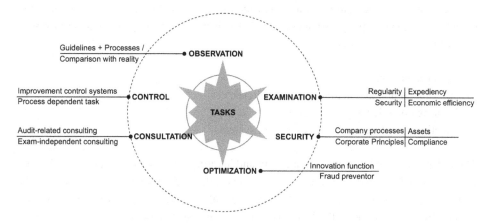

Fig. 1.7 Tasks of internal auditing

According to the principles of the DIIR, internal auditing, as shown in Fig. 1.6, comprises auditing and monitoring, advising, controlling, safeguarding, optimising and observing.[41]

As an instrument of corporate management, internal auditing assumes one of the essential tasks of corporate management in the form of auditing.[42] The audit is also the most classic task of the internal audit. It monitors the company processes, systems and decisions of the company management independently of the processes – i.e. with an external view.[43]

Numerous tasks that internal audit has to perform can be derived from the functions. Figure 1.7 shows examples of typical tasks assigned to the functions.

During its monitoring and auditing activities, Internal Audit gains a comprehensive insight and diverse knowledge of operational procedures, systems and processes. Likewise,

---

[41] DIIR – Deutsches Institut für Interne Revision e. V (2010, p. 9).

[42] Bünis and Gossens (2016, p. 15).

[43] Berwanger and Kullmann (2012, p. 49).

Internal Audit is very strongly networked within the company due to the numerous audit fields in which it is deployed. As a result, and due to its independence and objectivity, Internal Audit increasingly acts as an advisor to management.[44]

In addition to monitoring and advising, Internal Audit is also involved in the control of internal processes and organisational structures with regard to regularity and efficiency. In this context, control in the actual sense is more of a process-dependent task and thus does not actually correspond to the self-image of Internal Audit.

However, internal auditing supports corporate management and measures, evaluates and improves the effectiveness of internal control systems. Therefore, the control task of the internal audit is rather the support of the control activities of the company management. In addition, Internal Audit monitors the implementation of findings and recommendations from an audit as part of a follow-up audit, if required.[45]

The monitoring and auditing function of Internal Audit is always derived from the audit objectives of the company management. These always serve the general security objectives of the company, which would be

- Asset protection,
- Ensuring compliance with company policies,
- Securing compliance guidelines and
- Effectiveness of business processes and systems.[46]

The independent and objective monitoring and auditing of corporate processes and structures are supplemented by a future-oriented and efficiency-oriented optimization of the entire company. In this way, Internal Audit also makes a decisive contribution to the success of the company and to the fulfilment of the company's objectives.[47]

The fact that Internal Audit, as an autonomous, independent and objective functional unit, reports directly to the company's management and is not integrated into the company's processes, means that Internal Audit can also assume the role of an observer over the years and thus have an early and preventive effect on undesirable developments in the implementation of company processes, control systems and corporate objectives.

According to the IIA definition, the creation of added value for the entire company is one of the primary objectives of internal auditing, along with risk analysis and opportunity assessment.[48]

Depending on the focus, however, the tests are also carried out according to the following test criteria (cf. Fig. 1.8):

---

[44] Bünis and Gossens (2016, pp. 16–17).
[45] Bünis and Gossens (2016, p. 52).
[46] Bünis and Gossens (2016, p. 16).
[47] Otremba (2016, p. 92).
[48] Westhausen (2016, pp. 21–22).

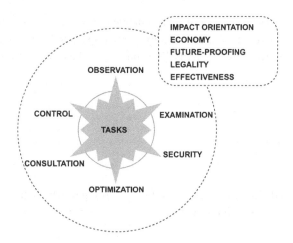

Fig. 1.8 Audit criteria for internal auditing

- Legality,
- Propriety,
- Security,
- Profitability,
- Future proofing,
- Ease of use/effectiveness, and
- Impact orientation.

In doing so, Internal Audit examines the lawful observance of corporate principles, strategies and guidelines and compliance with legal, supervisory and internal regulations, as well as the regularity and reliability of internal corporate processes such as accounting.

From a security perspective, Internal Audit performs a risk analysis, examines and evaluates the effectiveness of risk management, assesses opportunities, reviews the effectiveness of internal control systems and identifies irregularities and criminal acts.

The internal audit examines the efficiency and improves the operational activities and thus creates added value for the company.

Internal Audit works in a forward-looking manner and its work safeguards the company's assets and the quality requirements of the various stakeholders. It reviews the company's corporate and quality objectives for expediency, uncovers ineffective areas and improves the effectiveness and efficiency of management processes, monitoring processes, risk management processes and control processes.

Internal Audit works in an impact-oriented manner by systematically aligning its activities to achieve a positive impact and monitoring any impact and making it transparent.

Another rule prescribed by the IIA/DIIR as binding is the Code of Ethics, consisting of four principles

1. Righteousness,
2. Objectivity,
3. Confidentiality and
4. Expertise.

The Code of Ethics defines the rules of conduct according to which internal auditors must act.[49] Accordingly, it must exhibit the following characteristics:

- Integer,
- Confidential,
- Independent,
- Objective,
- Systematically,
- Purposeful,
- Process-independent,
- Competent and
- Risk-oriented.

For the result of the audit and monitoring of the internal audit to create added value for the company, the activities must be systematic and targeted.[50]

Due to the precise definition and the binding standards, a step-by-step process for internal auditing has developed in practice, consisting of four phases (cf. Fig. 1.9).

This model takes into account and combines all the principles, tasks and objectives described above.[51] At this point, reference is made to the detailed explanations on the process of an audit in Chap. 2.

Planning   Execusion   Reporting   Result checking

Fig. 1.9 Internal audit's regular procedure

---

[49] Otremba (2016, p. 96).
[50] Eulerich (2018, pp. 8–9).
[51] Otremba (2016, p. 97).

An internal audit supports the perception of the company management, it takes into account the transparency requirement and promotes "corporate hygiene". In addition, it serves prevention, as all procedures, processes and decisions are accessible to it and are thus "audit-threatened".

A company's management bears a considerable share of the responsibility if it decides against this instrument despite recognisable risks and if irregularities, breakdowns or cases of corruption occur as a result.

Internal Audit independently and objectively audits corporate processes and control systems on behalf of the company's management. Internal Audit performs these tasks in all areas of the company – commercial, technical and legal.[52]

A large part of internal auditing concerns the commercial affairs of a company, but technical matters are also becoming increasingly important.[53]

1.1.2   Technical Auditing

Internal audit can be organized on a functional level. Thus, specialized departments can be formed that cover the specialized tasks in addition to the classic commercial tasks. Depending on their professional training or expertise, the auditors can be deployed in the corresponding departments, e.g. commercial, IT and technical auditing (cf. Fig. 1.10).

Technical auditing focuses on the processes in the technical area of investments and maintenance.[54] It identifies and eliminates weaknesses and thus contributes to minimizing and preventing damage.[55]

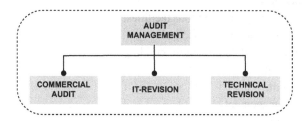

Fig. 1.10 Classification of technical auditing

---

[52] Berwanger and Kullmann (2012, p. 9).

[53] Berwanger and Kullmann (2012, p. 18).

[54] DIIR – Deutsches Institut für Interne Revision e. V (2010, p. 9).

[55] DIIR – Deutsches Institut für Interne Revision e. V (2010, p. 9).

In many companies, construction services are an essential part of the business processes, especially in the area of investments and maintenance. In order to cover these audit areas competently in the sense of the IIA/DIIR standards, special expertise is required.

The DIIR working group "Technical Auditing" was founded in 1980 because the DIIR recognized that technical processes were becoming increasingly important and therefore had to be included more extensively in the areas of responsibility of internal auditing. In accordance with the IIA/DIIR Principles for Technical Auditing, the working group has defined the following functions and tasks[56]

- Support,
- Advice,
- Control,
- Fuse and
- Observe.

Due to the ever-increasing number of topics, the working group was renamed "Construction, Operation and Maintenance" in 2006. The task of the working group is to develop working aids and guidelines for the tasks of technical auditing.[57]

The DIIR working group "Construction, Operation and Maintenance" has developed several different comprehensive audit concepts for technical auditing within the framework of internal auditing and has thus produced a practical guide for auditors.[58]

The DIIR Working Group Technical Auditing has identified the following audit areas of technical auditing:[59]

- Construction Services,
- Maintenance of buildings and outdoor facilities,
- Architectural and engineering services,
- Maintenance of technical equipment in buildings,
- Project Management,
- Claim Management,
- Environmental Management,
- Facilities Management,
- Dolose acts.

---

[56] DIIR – Deutsches Institut für Interne Revision e. V (2010, p. 5).
[57] DIIR – Deutsches Institut für Interne Revision e. V (2020, p. 1).
[58] DIIR – Deutsches Institut für Interne Revision e. V (2000, p. 64).
[59] DIIR – Deutsches Institut für Interne Revision e. V (2010, p. 5).

In numerous construction projects, the construction services are increasingly afflicted with defects and susceptible to manipulation. Technical auditing monitors and reviews all processes in a construction project, identifies the risks and potential for improvement.[60]

The audit of architectural and engineering services is one of the tasks within the scope of internal auditing. It should be carried out by the technical auditing department due to the required expertise and specialist knowledge.

If findings from the audit are already to be taken into account in a construction project under investigation, the audit should be started as early as possible and should be carried out in parallel.[61]

During execution, addenda are regularly issued in construction projects to record and enforce changes and additions to an existing contract. This can lead to significant cost increases in construction projects. A distinction must be made here as to whether claims are justified or not.

Justified claims can arise, for example, due to change requests by the client that were only formulated after the contract was concluded. Unjustified claims are often formulated where contractors have not correctly assessed the scope or complexity of a construction task when submitting a bid and now want to agree better prices afterwards.

The Technical Audit Department reviews and evaluates the tender, the contract and the project management for avoidable errors, risks and potential for improvement.[62]

If vicarious agents or employees of the client fail to scrupulously check a contractor's claims, corruption may be the reason. Corruption in the construction industry prevents competition and makes construction projects much more expensive. Project execution in particular harbours numerous risks for fraudulent acts such as

- Manipulation (e.g., through kick-back agreements),
- Price fixing (e.g. by cartelisation),
- Irregularities in the workmanship (e.g. due to the use of low-quality materials), and
- Incorrect invoicing (e.g. by disregarding the invoicing rules of the VOB/C).

The Technical Audit Department monitors and controls the construction processes and uncovers fraudulent activities.[63]

Facility management deals with the entire life cycle of buildings and outdoor facilities, from planning and use to deconstruction. Undetected problems, weak points or inefficiencies have a negative impact on the costs and thus on the asset protection of a company. An audit is of particular importance and is carried out by the technical auditing department due to the special knowledge required.[64]

---

[60] DIIR – Deutsches Institut für Interne Revision e. V (2010, p. 57).

[61] DIIR – Deutsches Institut für Interne Revision e. V (2000, p. 9).

[62] DIIR – Deutsches Institut für Interne Revision e. V (2017, backcover).

[63] DIIR – Deutsches Institut für Interne Revision e. V (2015, p. 57).

[64] DIIR – Deutsches Institut für Interne Revision e. V (2006).

Due to rising costs, optimizing the use and maintenance of assets is of particular importance to companies. Technical auditing assesses the quality of maintenance and identifies
optimization potential in the area of process and risk management.[65]

1.1.3   Construction Auditing

As a special audit with a focus on construction projects, construction auditing is a sub-area
of technical auditing and thus of internal auditing (see Fig. 1.11). The international standards and regulations of the IIA and, in Germany, of the DIIR apply to construction auditing in the same way as to all other sub-areas of internal auditing.

However, the auditing and consulting services always take place within the framework
of construction projects and with a focus on the construction activity. Depending on the
objective, it therefore includes a technical, economic and/or legal review of construction
procedures and construction processes. Due to the necessary interdisciplinary knowledge,
however, construction auditing is a little examined subarea within the framework of technical auditing or internal auditing.

For the construction companies there are numerous reasons to have their construction
projects checked by the construction audit:

- Profitability and cost pressure,
- Controling and monitoring services by the client due to outsourced functions and
  responsibilities,
- Identifying and eliminating risks,
- Exploiting potential for improvement and optimisation,
- Audit-proof implementation of construction projects,
- Combating fraudulent activities such as economic crime and corruption.

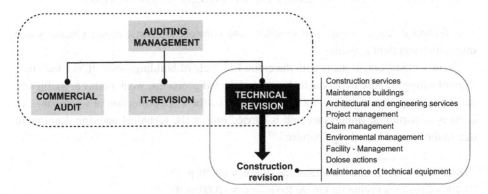

Fig. 1.11   Classification of construction auditing

---

[65] DIIR – Deutsches Institut für Interne Revision e. V (2013, 81).

Construction projects usually require a high capital outlay and are subject to extreme cost pressure. Ineffective processes, errors and defects mean that costs can increase uncontrollably. It is therefore important to identify cost risks in good time and to increase the efficiency of construction projects in order to control costs.[66]

Today, very few builders have all the tasks, processes and functions in a construction project in their own hands. Building owners delegate numerous tasks within the various departments in their company or to third parties outside the company. This means that duties and responsibilities are handed over and it is the client's task to coordinate and monitor all parties involved within a complex project execution.[67]

There are many types of risk in a construction project that are not limited to cost performance. In addition to budget risk and the risk of cost overruns, there are, to name but a few, for example, a

- Design risk due to incomplete tender documents,
- Construction schedule risk including construction schedule delays,
- Quality assurance risk due to numerous construction defects,
- Safety risk of the construction workers carrying out the work,
- Guarantee risk.

For non-construction companies and clients for whom construction is not part of their core business, such as manufacturing companies or trading companies, construction projects often represent an increased risk. Due to a lack of construction knowledge and experience, these builders are more easily duped than institutional builders.

For these companies, an appropriate internal control system and risk management is of particular importance.[68] In order to work successfully and efficiently, it is necessary to analyse these risks and to recognise and exploit potential for improvement in accordance with Fig. 1.12.[69]

Building owners who do not work with their own capital but, for example, with taxpayers' money or with the money of capital investors, are particularly required to act responsibly. For this reason, public clients in particular have to carry out, account for and document construction measures in an "audit-proof" manner. This means that the construction measures must be able to withstand an audit by the state or federal audit office.

The Land audit offices audit the budgetary and economic management of the Länder, while the Federal Audit Office audits the Federal Government. All measures that may have a financial impact are audited, with the scope of the audits themselves being determined.

---

[66] Risner (2012a, p. 43).
[67] Schwager and Fischer (2008, pp. 4–5).
[68] Schwager and Fischer (2008, p. 4).
[69] Risner (2012a, pp. 44–45).

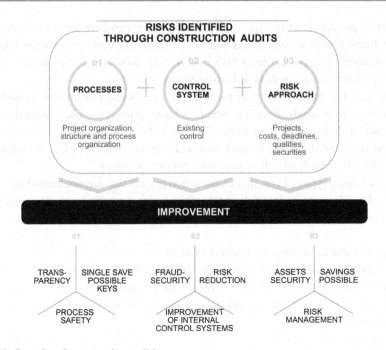

Fig. 1.12 Benefits of construction auditing

The audit shall be carried out in accordance with the standards of economy and regularity and shall cover compliance with the rules and principles applicable to budgetary and economic management. The audit results are presented annually in an audit report.

If the internal control systems or risk management systems are too weakly developed, there is a great risk of criminal acts to the detriment of the building owner. It is therefore the responsibility of every building owner to detect and eliminate such fraudulent acts in the construction sector and to implement preventive measures for the future.[70]

In addition to many other fraudulent acts in the context of white-collar crime, corruption in construction projects – not only abroad but also in Germany – is still widespread. Often, non-construction clients are susceptible to corruption and bribery due to their limited experience with construction projects and the lack of state supervision.[71]

---

[70] Schwager and Fischer, pp. 11–12.

[71] Transparency International e. V. (2005, p. 68).

The complexity of large construction projects in terms of organisation, technology and logistics offer favourable boundary conditions for corruption in construction:

- The large number of contractual relationships and participants, with a barely manageable number of payment channels, quickly leads to a loss of control. This circumstance makes it easy to hide bribes. The whereabouts of individual amounts is difficult to trace.
- The uniqueness of construction projects with regard to the design and the boundary conditions of the production makes it difficult to compare the specifications and subsequently the costs with those of other construction projects. This opens up "design scope".
- Opaque and difficult-to-audit procedures for awarding, executing and accounting for contracts make construction projects vulnerable to manipulation and fraud.
- Processes in the construction sector are not described at all or only inadequately and are therefore correspondingly unsuitable as a basis for an internal control system.

For these reasons, it makes sense for the client to commission the construction auditing department to audit, monitor and advise on the entire construction process from project development to handover. Particular attention is paid to the internal control systems, which are intended to ensure that the construction processes run economically and properly.[72]

The construction audit uncovers weaknesses and risks and derives potential for improvement from them. It helps to implement the improvements in the current project or to firmly implement them in the company processes for future projects.

Construction auditing ensures that construction projects comply with legal requirements and structural and technical standards and regulations.

Construction auditing also provides preventive protection against fraudulent acts. If the other party knows that specialists with a construction audit are involved in a construction project, the probability of detecting fraudulent acts is greatly increased and the construction project is therefore no longer so vulnerable.[73]

Finally, construction auditing helps to achieve the economic efficiency of construction projects and to prevent or minimise financial losses. From a construction point of view, construction auditing thus supports the achievement of overriding corporate and quality goals and ensures that construction projects meet the requirements of the "interested parties" in accordance with IIA/DIIR.[74]

In summary, the following functions and tasks result for the construction auditing:

- Detection and elimination of vulnerabilities,
- Damage prevention and mitigation,
- Create added value,
- Avoid additional costs,

---

[72] Schwager and Fischer (2008, pp. 5–6).

[73] Schwager and Fischer (2008, p. 13).

[74] Wingsch (2005, p. 66).

- Support companies in achieving their goals,
- Investigate and improve processes,
- Combating white-collar crime,
- Quality Assurance,
- Quality improvement,
- Efficient use of resources,
- Earnings improvement.

According to the IIA/DIIR Standards for Internal Auditing, a construction audit must also be independent and objective. The persons entrusted with the construction audit must not be part of the construction process, such as project managers, construction managers, engineers, architects.[75] Likewise, the tasks may not be assigned to a contractor who at the same time performs or has already performed other services in the construction project.[76]

Construction auditing requires special requirements compared to internal auditing, which is mainly based on economic knowledge. In addition to business knowledge, engineering knowledge is also required. The persons commissioned require a combination of construction engineering and commercial expertise and practical experience gained through construction projects.[77] This extensive knowledge makes it possible to recognize and evaluate the many pitfalls that can arise during the course of a construction project (see Fig. 1.13 for an example).

This requirement is based on the numerous tasks of construction auditing, which are specific to construction and are not found in this form in other sectors of the economy.

An inspection carried out by the construction audit is routine in many cases and is carried out at regular intervals. It is then carried out without any concrete suspicion of fraudulent, criminal or civil actions.

In the case of an ex-post review, the review is usually carried out retrospectively with greater effort. The main focus is on the subsequent determination of compliance and damage minimization. Ex-post inspections are therefore unsuitable as a control tool for an ongoing construction project. Measures for improvement or even optimization are then possible for subsequent construction measures.

During construction, on the other hand, an inspection can reveal problems in good time and measures can be initiated to eliminate them.[78] An ex-ante audit results from the current project status. In this case, the construction review becomes active in individual subsections and can intervene in the construction process in an advisory and corrective capacity.[79] Construction auditing is therefore most efficient during construction. However, care must be taken to ensure independence and objectivity. Construction auditing must not become

---

[75] Risner (2012a, p. 70).
[76] Wingsch (2005, p. 66).
[77] DIIR – Deutsches Institut für Interne Revision e. V (2010, p. 10).
[78] DIIR – Deutsches Institut für Interne Revision e. V (2010, p. 10).
[79] Schwager and Fischer, p. 6.

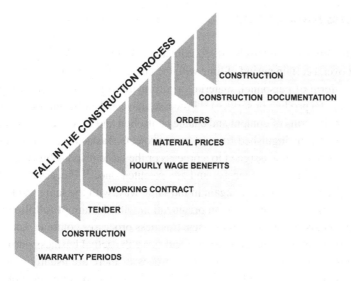

Fig. 1.13 Pitfalls of building as a test object

part of project management.[80] However, construction auditing is often called in on an ad hoc basis, i.e. in response to a specific initial suspicion.

The construction audit examines the entire construction process from project development to handover. The process can be audited as a whole or in individual sub-steps.[81]

Holistic testing has the advantage of being able to consider the entire process with all its dependencies and interconnections. However, a holistic inspection is very time-consuming and cost-intensive. As a rule, quick results that result in process improvements cannot be expected. The examination of partial areas, on the other hand, is faster, more flexible and more result-oriented. However, interfaces to other areas cannot usually be recorded.[82]

Audit is performed as a control review in which all processes, procedures, policies, practices, and internal controls are reviewed and monitored for security, propriety, risk, appropriateness, and efficiency. In cost control, the direct costs and the economic efficiency of a construction project are audited. Audit areas and audit subjects are explained in Sect. 2.3.

---

[80] Wingsch (2005, p. 66).

[81] Schwager and Fischer (2008, pp. 7–8).

[82] Wingsch (2005, p. 66).

1.2   Process Management

A process is the totality of interrelated, coordinated activities, tasks and procedures that are carried out over a certain period of time with the aim of transforming an input (an input or trigger in the form of personnel, material or information) into an output (an output or result in the form of a product or service).[83] Processes have a clear beginning and a defined end, are complete in terms of content and can be controlled and described in their entirety.

Processes must be distinguished from tasks or projects. Tasks are either the execution of a single activity that is not assigned to a process or the basic responsibility for a subject area. Projects are usually very extensive and are executed once.[84] Processes, on the other hand, can be repeated over and over again in the same way, with the same result.

Processes that run in a company or an organization with the aim of fulfilling the company and quality goals are business processes. Business processes combine work and production processes from different departments and serve the actual business purpose.

At the beginning and at the end of business processes there are usually customers or other company stakeholders who have certain expectations and requirements of the business process.[85] Companies that use business processes to organize the achievement of their goals are evidence of a high level of standardization and transparency of the processes in their company.

Business processes are organized across departments and hierarchies. Business processes influence the qualitative result of the company, profitability, customer and employee satisfaction as well as cost and time factors. For a company to be successful, the interaction of individual business processes and process steps must therefore be perfectly coordinated. This requires an overall entrepreneurial responsibility of the company management in the form of a process management for the planning, control and monitoring of business processes.[86]

The implementation of a process management allows the goal-oriented organization and control of business processes to achieve and optimize the company and quality goals.[87]

Process management can contribute to

- Achieve company and quality goals better and faster,
- increase efficiency, transparency, flexibility and quality,
- Reduce costs and time,
- Adapt to the requirements of the customers and to meet them better,
- Open up new business models,
- Organize and optimize business processes effectively and efficiently, and
- Comply with external legal standards as well as internal rules, orders and instructions.

---

[83] Schmidt (2012, p. 1).
[84] PICTURE GmbH (2020, p. 13).
[85] Seidlmeier (2019, p. 7).
[86] Schmidt (2012, p. 5).
[87] Schleinzer (2014, p. 33).

DIN ISO 9000 is an international series of standards that sets out the principles for action by companies to implement a quality management system. It promotes the choice of a process-oriented approach for the development, realization and improvement of the effectiveness of a quality management system. The standards describe the principles and terminology for quality management systems, define the requirements for quality management systems and provide a guide to effectiveness and efficiency.

DIN EN ISO 9000 assumes that a company consists of mutually influencing business processes and has established a process-oriented model for control, which is based on the PDCA method by Edward Deming. PDCA stands for Plan-Do-Check-Act or also Plan-Execute-Check-Act (cf. Fig. 1.14).

The PDCA method understands "planning" as the definition of processes that are in accordance with the quality policy and the quality objectives of the company. The "Execute" step involves the actual implementation of the processes. In the "Check" step, the processes and the resulting products are checked and monitored against the quality

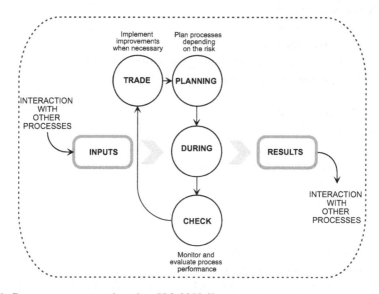

Fig. 1.14 Process management based on ISO 9000 ff

objectives. In the "Act" step, measures for improvement and optimization are taken on the basis of the inspection results.[88]

The standard aims to guide companies to implement process-oriented quality management systems. The business processes, which are aligned with the corporate objectives, the quality policy and the quality objectives, are in turn organized, controlled, evaluated and optimized with the aid of a process management system.

Numerous and diverse methods, tools and models are available to companies for this purpose, of which only a few can be mentioned here:

- KAIZEN – systematic learning processes with a collective approach,
- CIP – continuous improvement process,
- Benchmarking – comparison of own key figures with those of competitors,
- Lean Management,
- Change Management,
- Total Quality Management (TQM) – Reducing lead times while maintaining quality standards,
- Activity-Based Costing (PCR),
- Process Maturity Analysis,
- Risk Modality Analysis (RMEA),
- SWOT analysis – Strengths, Weakness, Opportunities, Threads,
- MITO – Management, Input, Transformation, Output Analysis,
- Business Process Reengineering,
- Customer Relationship Management,
- Risk Analysis.

These and numerous other methods are used in all areas of process management. All methods have the purpose:

- Optimize processes and make them more effective and efficient in order to minimize costs,
- Link and synchronize processes to shorten execution times
- Improve the quality of the output of processes,
- Automate and standardize processes and activities to avoid errors, reduce costs and shorten times,
- Identify potential for outsourcing processes in order to better focus on core competencies as a company.

[88] DIN EN ISO 90001, pp. 7–8.

1.2.1 Process Analysis

In order to be successful on the global market and to meet the expectations of customers, a company must permanently review and adapt its entire corporate and quality policy as well as its quality objectives. This means that the processes in a company are also subject to constant change.

Within the framework of a process management system, all tasks from the identification of processes to the improvement or even optimization of processes are performed. This includes, as shown in Fig. 1.15:

- Process organization,
- Process control,
- Process design,
- Process controlling and analysis and
- Process Optimization.

In order to obtain a precise overview and a comprehensive understanding of a process, with the aim of identifying weak points and risks and recognizing optimization potential, a systematic examination of processes and the individual activities and tasks contained therein is required. For this purpose, business processes within a company are documented, analyzed, evaluated and optimized. Thus, process analysis is a step in process management of great importance.

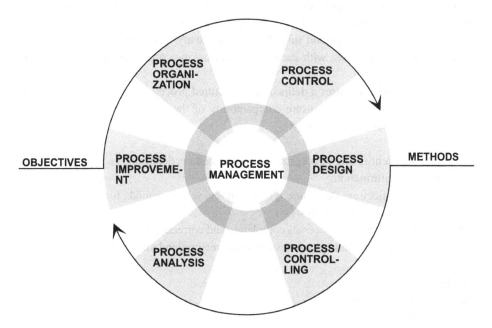

Fig. 1.15 Tasks of process management

Process analysis is used to systematically determine the current state of a process and compare it with the target state. The process can thus be checked for the achievement of the set goals and possible deviations can be evaluated.[89]

This results in starting points for process control, improvement and optimization.[90] A process analysis also serves to make interconnected processes more transparent. In this way, interfaces between processes and to other company units can be checked.

Another aspect is the possibility to check with an analysis of the processes whether they are also aligned with the corporate goal and the quality policy. In this way, the entire process management is aligned with the competitiveness and success of a company and the business result is increased through an improvement in effectiveness and efficiency.

A process analysis considers and evaluates business processes, sub-processes and activities from different aspects[91] and provides answers to possible questions:

- Why do the results not match the expectations?
- Why are the goals not being met?
- How can processes be improved?
- Where do new processes need to be developed?
- Which processes need to be synchronized with each other?
- How can processes be made more efficient with the help of key figures?

The process analysis can be performed as a top-down analysis or as a buttom-up analysis. In the top-down analysis, the processes are evaluated starting from the highest company level with regard to the superordinate quality objectives. Various business processes are broken down to sub-processes and individual process steps. The bottom-up analysis starts with individual process steps and sub-processes and analyzes the effectiveness and interfaces of these sub-processes with each other and how they are ultimately interlinked in business processes.

A process analysis requires a defined and structured procedure and usually uses standardized analysis methods to ensure comparability of the analyses. A process analysis usually runs in the following phases:

- Preparation: Identify and select process, plan procedure, determine analysis criteria and gather initial information,
- Information collection: determine type of survey, select methods for analysis, collect information and conduct analysis,
- Documentation: Present processes completely and correctly, map analysis evaluation in a documentation, evaluate processes and make recommendations for action.[92]

---

[89] Seidlmeier (2019, p. 113).
[90] Matschechin (2017, p. 6).
[91] Hirzel et al. (2013, p. 19).
[92] Matschechin (2017, p. 9).

In order to carry out a process analysis, criteria must be defined in advance according to which the analysis is to be carried out. If the focus is on evaluating the processes with regard to the company's policy and quality objectives, these include higher-level objectives such as quality, flexibility, time and costs.[93]

If, on the other hand, the focus is on improving sub-processes, criteria such as the use of employees and resources or information policy are in the foreground.

1.2.2   Maturity Assessment

The improvement of processes is one of the most important tasks in process management. Before optimizing company processes, the focus is first on analyzing the processes. In doing so, the question of the development status of the examined processes should be answered.[94]

For quality management, the DIN EN ISO 9001 standard requires the introduction of certain processes for the operational procedures to achieve the company and quality objectives. Within the scope of an audit, it is checked whether these processes are carried out as planned in the company. As a result, the company is certified that the actual processes correspond to the planned processes.

Maturity models have a similar approach. They use objective criteria to determine the performance of existing processes and thus the performance of companies in relation to corporate goals.[95] They are used to determine and objectively assess the current state of development, to compare the performance of different processes and to identify potential for improvement.[96]

A development process is determined for selected processes. This determines the conditions under which a process has reached the "optimum" degree of maturity.[97] Maturity refers to the achievement of a predefined state of a thing or an object.[98]

Maturity models comprise several development stages – maturity levels – that build on each other. For each development stage, starting at the initial stage and ending at optimal maturity, there are predetermined criteria in various categories that must be met by the process.[99]

Maturity models examine many characteristics, properties, and objectives of the processes under study. These describe the categories or dimensions to be examined. Dimensions are measurable or describable characteristics of a process. For each

---

[93] Schmidt (2012, p. 5).

[94] Röglinger and Kamprath (2012, p. 2).

[95] Kübel (2013, p. 72).

[96] Röglinger and Kamprath (2012, p. 4).

[97] Röglinger and Kamprath (2012, p. 2).

[98] Schleinzer (2014, p. 5).

[99] Röglinger and Kamprath (2012, p. 4).

dimension under study, a process can reach different levels of maturity. This results in a multi-dimensional differentiated analysis of the process.

In the maturity model, it is determined which process is to be examined, according to which criteria this is to be done and with which parameters the criteria are to be evaluated.[100]

The maturity level to be determined is defined by certain characteristics and characteristic values to be achieved. The better the defined criteria are met, the higher the maturity level. A maturity level is reached when both the requirements applicable there and the requirements for the previous level are fulfilled.[101]

Data collection for the application of maturity models can be done using different tools and methods (observation, interviews, questionnaires). Whichever model is to be used, data about the process must be collected and evaluated at a certain point in time. This allows the actual situation of the process to be examined and recommendations for action to be derived.[102]

In the context of process management, a distinction is made between maturity models that focus on the individual process on the one hand and on the entire company-wide process management on the other.[103]

One of the first maturity models was developed in the field of quality management as early as 1979. The **Quality Management Maturity Grid** (QMMG) was developed to increase the quality of products through continuous improvement of manufacturing processes. The QMMG describes six dimensions with the five maturity levels "Uncertainty", "Awakening", "Realization", "Understanding" and "Safety".[104]

As another model, the **Capability Maturity Model Integration** (CMMI) is designed to determine an organization's ability to implement processes to achieve organizational goals and use the results to drive improvement efforts.

The CMMI goes back to the **Capability Maturity Model** (CMM) from 1987, with which the Software Engineering Institute (SEI) wanted to determine why some software manufacturers were more successful on the market than others.[105] The CMMI model describes the maturity level in the five stages "Initial", "Repeatable", "Defined", "Managed" and "Optimizing".[106]

Since then, around 150 maturity models have been developed for various areas, such as project management, process management, product development, human resources

---

[100] Jakoby (2019, p. 185).

[101] Schleinzer (2014, p. 5).

[102] Schleinzer (2014, p. 6).

[103] Röglinger and Kamprath (2012, p. 4).

[104] Kübel (2013, p. 69).

[105] Jakoby (2019, pp. 185–186).

[106] Kübel (2013, p. 69).

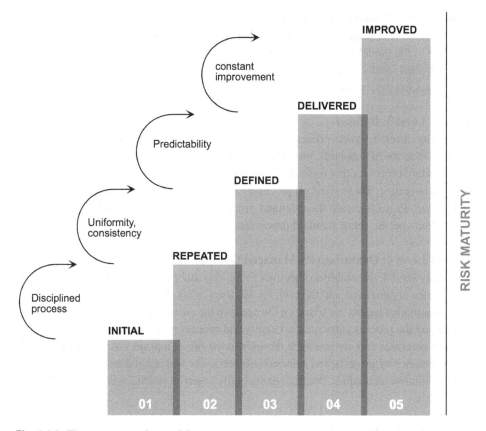

Fig. 1.16 Five-stage maturity model

management, information technology, etc.[107] The Capability Maturity Model Integration (CMMI) is the most frequently used model.[108]

In the application, for example, a 5-level maturity model can be applied. The five maturity levels, each of which represents a basis for further process improvement, are numbered from 1 to 5 (cf. Fig. 1.16).

Maturity Level 1: Initial
At maturity level 1, workflows are usually ad hoc and chaotic. The organization usually does not provide a stable environment to support the workflows. Success in such organizations depends on the competence and commitment of the employees rather than on the use of a proven process.

---

[107] Kübel (2013, pp. 69–70).
[108] Röglinger and Kamprath (2012, p. 2).

Maturity Level 2: Guided/Repeatable
At maturity level 2, projects have ensured that work processes are planned and executed according to the guidelines. Professionals with sufficient resources are deployed to produce controlled results, workflows are monitored, controlled and audited, and compliance with the process description is assessed.

Maturity Level 3: Defined
At maturity level 3, work processes are well characterized and understood and are described in terms of standards, procedures, tools and methods. The organization-specific set of standard processes that forms the basis for maturity level 3 has been established and improved over time. These standard processes are used to establish consistency within the organization. Projects create their defined processes using tailoring guidelines from the organization-specific set of standard processes.

Maturity Level 4: Quantitatively Managed/Guided
At maturity level 4, quantitative objectives for quality and process performance are established for the organization and the projects and used as criteria for managing the projects. These quantitative targets are based on the needs of the customers, the end users, the organization and the process participants. Quality and process performance are understood as statistical measures and are managed throughout the life of a project. At maturity level 4, the performance of projects and selected sub-processes is managed using statistical and other quantitative techniques. Predictions are partly based on statistical analysis of detailed process data.

Maturity Level 5: Optimizing
At maturity level 5, an organization continuously improves its processes based on a quantitative understanding of its business goals and performance needs. The organization uses a quantitative approach to understand the inherent variation in the process and the causes of process outcomes.

The qualitative classification of quality and risk of a process to the maturity levels is summarized in Table 1.1.

The evaluation of the qualitative findings from the review questions that are examined in the course of a building review can be used to assess the degree of maturity. This can be

Table 1.1 Qualitative classification of quality and risk to maturity levels

Stage	Description of the degree of maturity	Quality of the process	Risks of the process
1	Initial	Lowest	Highest
2	Guided	Low	High
3	Defines	Medium	Medium
4	Qualitatively managed	Higher	Low
5	Optimizing	Highest	Lowest

TIRE WHEEL	FREE OF OBJECTION	RECOMMENDA-TION	CHANGE REQUIREMENT
Level Description	★ ★ ★	★ ★	★
1 Initial	0%	0%	100%
2 Guided	5%	45%	50%
3 Defined	20%	60%	20%
4 Quantitatively guided	50%	45%	5%
5 Optimizing	80%	20%	0%

Fig. 1.17 Maturity levels as a function of test results

done, for example, by means of a quantitative allocation according to an evaluation matrix, as Fig. 1.17 shows.

An assignment to maturity level 1 (initial) is made if less than 5% of the questions were answered without objections and more than 50% of the questions revealed a need for change.

An assignment to maturity level 2 (Guided) is made if at least 5% of the questions were answered without objections and no more than 50% of the questions revealed a need for change.

An assignment to maturity level 3 (Defined) is made if at least 20% of the questions were answered without objections and no more than 20% of the questions resulted in a need for change.

An assignment to maturity level 4 (quantitatively managed) is made if at least 50% of the questions were answered without objections and no more than 5% of the questions revealed a need for change.

An assignment to maturity level 5 (optimizing) is made if at least 80% of the questions were answered without objections and less than 5% of the questions resulted in a need for change.

This is illustrated below using an example in which change and supplement management was reviewed. A total of 135 audit questions were considered for the fields of action examined. The results of these are shown in Fig. 1.18.

The overview in Fig. 1.18 shows that, especially in the audit field "forward-looking claim management", the statistical base is so small that the results can only be related to the individual case examined. General findings beyond the individual case can only be derived to a limited extent.

The percentage distribution of the response categories for each test field examined is shown in Fig. 1.19.

The diagram in Fig. 1.19 shows that only a small proportion of the company's **change** and **supplement management** can be rated as complaint-free. Only 17% of the audit questions did not lead to any objections. Since a need for change was also identified in

EXAMINATION FIELD	FREE OF OBJECTION	RECOMMENDA-TION	CHANGE REQUIREMENT
REVIEW OF CHANGE & SUPPLEMENT MANAGEMENT	Σ23 ★★★	Σ86 ★★	Σ26 ★
1 Documentation	5	26	11
2 Decision-making powers	2	11	0
3 Approach to Subsequent Testing	10	18	9
4 Examination basis of claim	1	14	4
5 Examination claim level	1	9	1
6 Forward-looking claim management	4	8	1

Fig. 1.18 Example for the determination of the maturity level – step 1

EXAMINATION FIELD	FREE OF OBJECTION	RECOMMENDA-TION	CHANGE REQUIREMENT	TIRE WHEEL
REVIEW OF THE AMENDMENTS-& SUPPLEMENTARY MANAGEMENT	Σ23 ★★★	Σ86 ★★	Σ26 ★	2 Guided
1 Documentation	12%	62%	26%	2 Guided
2 Decision-making powers	15%	85%	0%	2 Guided
3 Approach Supplementary audit	27%	49%	24%	2 Guided
4 Examination claim basis	5%	74%	21%	2 Guided
5 Examination of claim amount	9%	82%	9%	2 Guided
6 Forward-looking Claim management	31%	62%	8%	3 Defined

Fig. 1.19 Example for the determination of the maturity level – step 2

19% of the audit questions, the overall maturity level is to be described as "**Managed**". There is a need for action on these points.

For the audit field "**Documentation**", the overall maturity level is to be described as "**Guided**". This rating results from the fact that only 12% of the audit questions were answered without objections. This percentage should not be less than 20% for a classification as a "Defined" process. In addition, a need for change was identified in 26% of the audit questions. This proportion should not be more than 20% for a classification as a "Defined" process.

For the examination field "**Decision-making competencies**", the overall maturity level is to be described as "**Guided**". This rating results from the fact that only 15% of the audit questions were answered without objections. This percentage should not be less than 20% for a classification as a "Defined" process. A positive aspect is that no urgent need for change was identified in this area of the audit.

For the audit field "**Approach of the supplementary audit**", the overall maturity level is to be described as "**Guided**". This assessment results from the fact that, with 27% of the audit questions, the proportion of complaint-free audits meets the required proportion of no less than 20%. On the other hand, a need for change was identified in 24% of the examination questions. This proportion should not be more than 20% for classification as a "defined" process.

For the examination field "**Examination of the basis for claims**", the overall maturity level is to be described as "**Guided**". This rating results from the fact that only 5% of the audit questions were answered without objections. This percentage should not be less than 20% for a classification as a "Defined" process. In addition, a need for change was identified in 21% of the audit questions. This proportion should not be more than 20% for a classification as a "Defined" process.

For the audit field "**Examination of the level of claims**", the overall maturity level is to be described as "**Guided**". This rating results from the fact that only 9% of the audit questions were answered without objections. This proportion should not be less than 20% for a classification as a "Defined" process. The share of 9% of the audit questions for which a need for change was identified is already sufficiently low for a higher classification.

For the audit field "**forward-looking claim management**", the overall maturity level can be described as "**defined**". 31% of the audit questions did not lead to any objections, while only 8% of the questions revealed a need for change.

1.2.3 Risk Management

Companies have to change and develop permanently because numerous framework conditions require this. These include:[109]

- Internal and external influences in ever-changing markets,
- A high degree of uncertainty due to constantly growing market pressure,
- Numerous competitors, and
- Numerous global challenges.

This is accompanied by numerous opportunities, but also risks.[110]

The implementation of corporate objectives and quality policy and the associated creation of added value for the various stakeholders is fraught with uncertainty and insecurity. According to the generally accepted understanding, uncertainty comprises on the one hand the risks but also the opportunities for companies in achieving their goals.[111]

---

[109] Eulerich (2018, p. 1).
[110] Eller (2010, p. 5).
[111] COSO – The Committee of Sponsoring Organizations of the Treadway Commission (2004, p. 1).

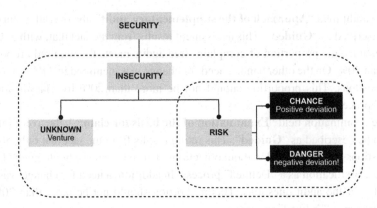

Fig. 1.20 Definition of risk

Every entrepreneurial activity involves taking advantage of opportunities, but also taking risks. Risk-free business management is not possible, but dealing with it is of crucial importance. The value of a company depends crucially on how it deals with the existing opportunities and risks.[112]

Every decision that a company has to make involves risks. In companies, these risks accompany all strategic and operational processes. Weighing up opportunities and risks helps to reduce costs, weigh up entrepreneurial decisions against each other and thus increase the success of the company.[113]

In common parlance, the term risk tends to have a negative connotation and means a danger that is associated with a disadvantage.[114] The international standard DIN EN ISO 31000, on the other hand, defines risk as the effect of uncertainty on objectives. Uncertainty arises from a lack of information, understanding or knowledge about an event.[115] Here, risk describes a situation whose outcome is fundamentally open.

In the literature, uncertainty is divided into uncertainty and risk (cf. Fig. 1.20). Uncertainty arises when the dependencies that may occur in a decision-making process or a process cannot be identified and assessed in more detail.[116] Risk, on the other hand, means an identifiable and assessable impact of a deviation from expectations. The probability of occurrence of an existing risk can be determined and recommendations for action can be derived from it.

---

[112] Brauweiler (2019, pp. 1–2).

[113] Gleissner (2016, p. 32).

[114] Hoffmann (2017, p. 1).

[115] DIN ISO 31000, pp. 8–9.

[116] Hoffmann (2017, p. 2).

Events can therefore have both negative and positive effects. If an event develops more negatively than expected, then this means a danger which must be identified, evaluated and improved. If an event develops more positively than expected, then it represents an opportunity to achieve corporate goals and increase value creation.[117]

In order to better identify and document risks, they are classified into different risk types. Risk types can be:

- Credit risk (deterioration in creditworthiness, insolvency, unwillingness to pay),
- Political risk (country risk, transfer risk, insurgency, terror, war),
- Market price change risk (commodity prices, interest rates, exchange rates),
- Operational risk (failure of systems, accidents, errors in processes),
- Legal risk (prohibitions, bids, regulation, lawsuits, contractual risks),
- Behavioral risk (dolose actions, opportunism, moral hazard, compliance),
- Reporting risks,
- Sales risk.

These can be classified according to their scope and objectives, taking into account the following aspects (cf. Fig. 1.21):

- Risk impact (costs, revenues, time, quality),
- Time of creation (planning, preparation, execution, operation),
- Type of risk (legal, financial, strategic, technical, operational),
- Risk carrier (management, employee, supplier, customer),
- Ability to influence (company-specific, global, systematic, unsystematic).[118]

Dealing with risks in a company should always be part of a comprehensive risk policy or risk strategy. The standard DIN EN ISO 31000 describes the risk policy in a company as the summary of the general intentions and orientations of a company in connection with risk management. The risk policy reflects the attitude of a company with regard to the assessment and acceptance or avoidance of risks.[119]

A sensible risk strategy creates risk awareness in the company and thus determines up to what point the opportunities and risks are taken into account in the implementation of the company's goals or at what point the company's goals may have to be adjusted because the risks are too great and the previous expectations cannot be met.[120]

The risk strategy defines and formulates objectives and framework conditions for dealing with risks at all levels of the company.[121] The company management thus establishes

---

[117] COSO – The Committee of Sponsoring Organizations of the Treadway Commission (2004, p. 2).
[118] Hoffmann (2017, p. 28).
[119] DIN ISO 31000, p. 9.
[120] Eller (2010, p. 34).
[121] Hoffmann (2017, p. 16).

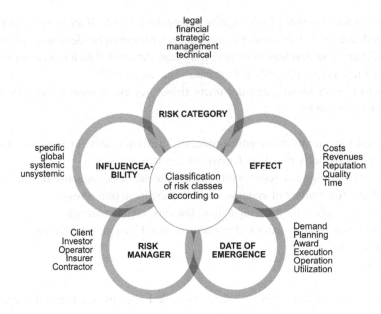

Fig. 1.21 Risk classes

the guidelines of the risk policy and defines guidelines in dealing with risks in terms of objectives, organization, communication, documentation, etc.[122]

Within the framework of the defined risk appetite, management determines the limits within which risks are accepted. The risk strategy also defines the environment in which the risk management system is implemented.[123]

Part of the risk strategy should also always be the consideration of the risk appetite. The risk appetite in a company says something about the attitude of the company management towards risk. It has an impact, for example, on the capital reserve in a company for hedging against risks, on the time period for which the potential risks are identified and up to which probability of occurrence risks are tolerated.[124]

Within the framework of the Three Lines of Defense model, the risk strategy is the starting point for the internal control system (ICS). The ICS is the internal control and

---

[122] Hoffmann (2017, p. 18).

[123] Brauweiler (2019, p. 3).

[124] Eller (2010, pp. 29–30).

monitoring system for all business processes and procedures in a company that are essential for the corporate policy and quality objectives.[125]

The ICS serves to monitor a company in relation to its strategy and objectives and, in addition to reviewing processes, also focuses on asset preservation and compliance with rules and guidelines. Risk management is only one component of the ICS, which also considers quality management, compliance management and process management in addition to risks.[126]

A systematic approach in the form of a risk management system is required to monitor and control risks in a targeted and comprehensive manner.[127]

With DIN EN ISO 31000, an international standard was adopted by the International Standardization Organization (ISO) in November 2009, in which the principles and guidelines for effective handling of corporate risks are anchored. DIN EN ISO 31000 recommends that all companies create, implement and continuously improve a framework for a risk management process. Risk management can be applied to the entire corporate structure in all areas and at all levels or only to specific projects and activities.[128]

The international standard DIN EN ISO 9001 on quality management also emphasises the importance of "risk-based thinking" in connection with the introduction, organisation and improvement of a quality management system.[129]

Alongside the COSO framework, DIN EN ISO 31000 is the world's leading standard for risk management.[130] The standard is intended to link existing management systems, such as quality management, with risk management. The aim is to consider, implement and improve risk management holistically at all levels and in all areas within the company.[131]

DIN EN ISO 31000 describes the principles for effective risk management at all levels of a company.[132] According to this, risk management is characterized by the fact that it

- creates and protects values,
- is an integral part of all organisational processes,
- is part of the decision-making process,
- explicitly addresses uncertainty,
- is systematic, structured and timely,
- is based on the best available information,
- is tailor-made,

---

[125] Schleinzer (2014, pp. 36–39).

[126] Brauweiler (2019, p. 1).

[127] Eller (2010, p. 28).

[128] DIN ISO 31000, p. 5.

[129] DIN EN ISO 90001, p. 8.

[130] ISACA Germany Chapter (2014, p. 2).

[131] Romeike (2018, p. 21).

[132] DIN ISO 31000, pp. 15–16.

- Human and cultural factors taken into account,
- is transparent and does not exclude,
- is dynamic and iterative,
- reacts to changes,
- facilitates the continuous improvement of the organization.

Another internationally significant framework is the Enterprise Risk Management – Integrated Framework, also known as COSO II for short, published in 2004 by the Committee of Sponsoring Organization of the Treadway Commission (COSO). In this framework, risk management is regarded as a process that is applied to identify events that are significant for the company's policy and to create security for the achievement of objectives.[133]

The COSO II model consists of a three-dimensional approach. The so-called COSO cube (cf. Fig. 1.22) takes into account the target categories, components and organisational aspects of the company in order to represent a company-wide risk management.[134]

The first aspect of the COSO risk management system takes into account the various short, medium and long-term goals of a company and focuses on their implementation and improvement.

The second aspect comprises the eight components of risk management, which also influence each other.

The third aspect focuses on the organization of the company and takes into account that risks can occur at any level and in any area and can influence the overall quality objectives.[135]

Well-implemented risk management increases the likelihood of companies achieving their own goals, creates an awareness of proactive risk management within the company, identifies opportunities and ameliorates threats, increases the confidence of the company's stakeholders, avoids and minimizes damage, and increases the company's resilience in global competition.[136]

The tasks of risk management include the treatment, control, monitoring and documentation of risks. It identifies certain risks and provides recommendations for dealing with them. This results in decision-making options for the company's management.[137]

Unforeseeable risks can be mastered by identifying, avoiding and reducing critical situations at an early stage. In this way, a well-functioning risk management system also acts as an early warning system for the company's management.[138]

---

[133] Schleinzer (2014, p. 18).
[134] Schleinzer (2014, p. 19).
[135] Otremba (2016, pp. 111–112).
[136] DIN ISO 31000, pp. 5–6.
[137] Hoffmann (2017, p. 4).
[138] Brauweiler (2019, p. 1).

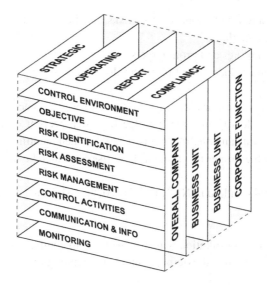

Fig. 1.22 COSO cube

Once the risks have been identified and assessed, risk management can ensure the sustainable existence and success of the company by permanently monitoring and improving them.[139]

Important goals of a well-functioning risk management are:[140]

- Improvement of information systems,
- Improvement of internal and external processes,
- Compliance with all legal requirements and internal rules and regulations,
- Creation of a high level of risk awareness among all company stakeholders,
- Increase transparency and avoid a "bad reputation" of the company,
- Ensure functioning transparent information systems,
- Development of a functioning early warning system and the early identification of opportunities and risks,

---

[139] Brauweiler (2019, p. 5).

[140] Brauweiler (2019, p. 2).

- Regular identification and assessment of risks and the consistent implementation of measures to eliminate them,
- Achievement of earnings and profit targets,
- Minimize costs and reduce loss of resources due to risk.

Risk management is process-oriented and consists of several successive steps.[141] Internal controls are used to identify in good time the risks that prevent or at least impede the achievement of corporate and quality targets. These are then identified, analyzed and evaluated. Measures for the proper handling of these risks can then be developed, introduced and monitored.[142]

In the international standard DIN EN ISO 31000, the risk management process consists of the following steps

- Communication and consultation with and between all company stakeholders,
- Establishing the internal and external context and defining the influencing factors and risk criteria,
- Risk assessment with the sub-tasks of risk identification, risk analysis and risk evaluation, risk management through the selection of suitable measures, as well as
- Monitoring, verification and documentation of the selected measures.[143]

In the COSO framework, the process steps are described as components in the second dimension of the COSO cube similar to the DIN EN ISO 31000 standard. These include:

- Internal Environment,
- Goal Setting,
- Event Identification,
- Risk Assessment,
- Risk management,
- Control Activities,
- Information and communication,
- Surveillance.[144]

As shown in Fig. 1.23, all risk management systems are a systematic application of processes to communicate, identify, analyse, assess, resolve and monitor potential risks.[145]

---

[141] Hoffmann (2017, p. 15).

[142] Berwanger and Kullmann (2012, p. 61).

[143] DIN ISO 31000, pp. 22–29.

[144] COSO – The Committee of Sponsoring Organizations of the Treadway Commission (2004, pp. 3–4).

[145] DIN ISO 31000, p. 10.

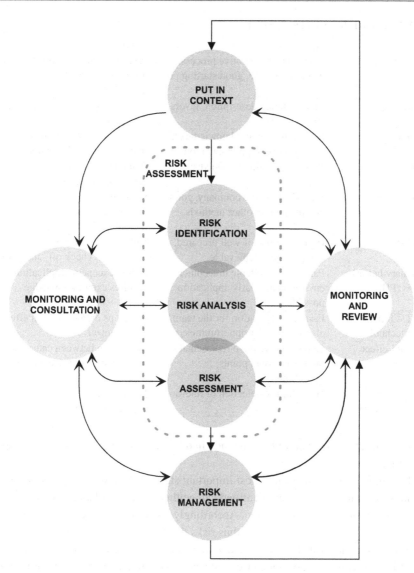

Fig. 1.23 Risk management

Risk identification describes the process of finding, recognizing and describing risks.[146] In this process, existing individual risks as well as possible future risks are identified in relation to the company's and quality objectives.[147] Risk identification should be carried

---

[146] DIN ISO 31000, p. 11.

[147] Schleinzer (2014, p. 27).

Table 1.2 Methods for risk identification

Method	Per	Contra
Brainstorming	High quantity, creative process, quick to execute, good starting point	Risk fields are overlooked, unstructured, prefer "alpha animals"
Brainwriting	More detailed description, favors completeness	More complicated to implement, unstructured
Interviews	Inclusion of expert knowledge, high quality, good supplementation	Costly, lengthy, resource-intensive, possibly inquisitorial
Document analysis	Securing existing knowledge	Elaborate, lengthy, backward-looking
Lessons learned	If available in the company, good complement to other methods	Requires an open corporate culture, not suitable for hierarchical structures
Checklists	High expertise, if well prepared, completeness achievable	Unknowns remain unnamed, encourage "check off" mentality
Preliminary hazard analysis (PHA)	Particularly suitable in technical environments, early application possible	Limited scope of application, requires expert knowledge
SWIFT – Structured what-if technology	Fast, no preparation for the team, also suitable for opportunities	Expert as organizer, high effort for organizer
Cause-and-effect analysis	Graphic descriptive representation, points out weak points, opportunities	Interaction between categories is suppressed, expertise required
Consequence/ probability matrix	Good for pre-selection of relevant risks	Error-prone, since two parameters have to be estimated

out systematically, in a structured manner and in relation to the previously defined aspects and criteria.[148]

Risk identification is one of the most important steps in the risk management process. After all, risks that have not been identified can neither be analyzed and evaluated further nor can appropriate measures be taken. Accordingly, identification should be repeated on a regular basis, as projects develop and thus risks that have been identified once no longer apply or new ones are added. Processes that are changed therefore require a new analysis.

A distinction is made between causes, indicators and effects. Causes are certain findings whose existence or subsequent occurrence could lead to risks. Indicators are indications that show the occurrence of a risk and effects are possible consequences that result if a risk occurs.

Several complementary methods can be used for identification (see Table 1.2). The methods range from simple questionnaires or workshops to complex survey systems, process and failure analysis tools.[149] It is necessary to agree on the advantages and

---

[148] Oepen (2012, p. 73).
[149] Hunziker and Meissner (2018, p. 100).

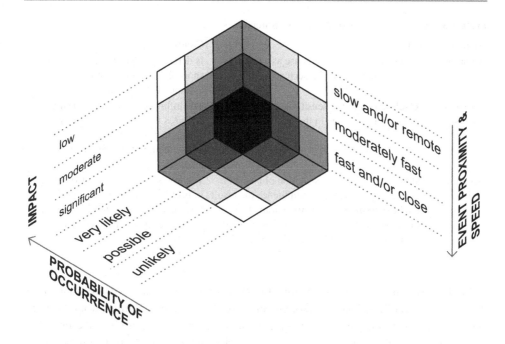

Fig. 1.24 Risk probability

disadvantages, strengths and weaknesses in relation to the particular situation where they are to be applied. Some of the methods used and their advantages and disadvantages are described below.

After the risks have been identified, they are analyzed in terms of their causes and sources.[150] In the risk analysis, the probability of occurrence is also important and whether the risk is a potential opportunity or danger, i.e. how high the damage resulting from the risk will be. This can be done with the help of a large number of qualitative and quantitative analysis tools.[151]

As a rule, the analysis is carried out with the aid of a diagram. In this diagram, the risk is assessed according to the probability of occurrence and the amount of the expected loss. As shown in Fig. 1.24, the time or speed at which the risk is to be expected can also be taken into account in the analysis.[152]

To estimate the probability of occurrence, frequencies of events in historical data can be used. This is only possible for events whose history is known, but not applicable for very rare events or events that have never occurred.

---

[150] Hoffmann (2017, pp. 29–32).

[151] Schleinzer (2014, p. 27).

[152] Brauweiler (2019, p. 8).

Table 1.3 Quantitative methods for risk measurement

Method	Per	Contra
Volatility	Widespread and common use as a risk measure in the financial sector	Is derived from history, scaled from short observation periods and assumed to be constant
Sensitivity analysis	Result indicates the sensitivity of the result to a minimal change in the risk factors again	The informative value of sensitivity analyses is very limited in most areas of application; it makes more sense to use them in the form of scenario analyses
Scenario analysis	Model is a good representation of a section of reality, parameters that influence the result can be determined perfectly; parameter variation can be derived from history; model is easy to understand, results are plausible	Too few scenarios to identify all impacts; history may not repeat; assumptions may not hold; selected section for model neglects risks; results without probability of occurrence

Well-known methods for predicting the probability of occurrence are Fault Tree Analysis or Event Tree Analysis. Both methods evaluate the probabilities from different angles. Fault Tree Analysis starts with a possible failure and determines the probable events and causes that can lead to this failure, and Event Tree Analysis determines what the possible reactions to an event can do for success or failure.

Another possibility is quantification on the basis of known statistical distributions. Quantitative methods require the calculation of probabilities of occurrence, possible damage and thus knowledge about the distribution of events.

Examples of quantitative methods for measuring risk are shown in Table 1.3.

Risk analysis is followed by risk assessment. In the DIN EN ISO 31000 standard, risk assessment is described as a process in which the results of the risk analysis are compared with the predefined criteria.[153] As with every process analysis, a target-performance comparison of the qualitative and quantitative risk criteria also takes place during risk assessment. After the assessment, the risks can be prioritized according to the criteria and the company's defined tolerances. A need for action in dealing with the identified and analyzed risks can then be derived from this.[154]

In doing so, the company must ask itself whether the deviation is tolerable or whether the risk should be dealt with. The cost-benefit effect also plays a role here. Ideally, limit ranges for classification are defined in advance.[155]

Risk assessment is always subject to a certain degree of uncertainty. On the one hand, this is due to the fact that there is no experience in the company for certain types of risk or that these are interpreted differently by the various company stakeholders. A certain

[153] DIN ISO 31000, p. 13.
[154] Schleinzer 2014, p. 27.
[155] Hoffmann (2017, p. 32).

amount of experience in assessing risks must therefore be assumed and all those involved must be aware of the uncertainties.[156]

The greatest challenge in the valuation process is that when individual risks accumulate, they influence each other qualitatively, quantitatively and in terms of time.

This can lead to risks being underestimated, e.g. because the expected values add up, or overestimated, because the damage potentials add up. Due to a lack of experience, false assumptions about mutual influence can occur, or risks with different consequences that seem to fit in terms of content accumulate.

In the literature, the measures following the identification, analysis and evaluation of risks are summarized under various terms. The COSO framework, for example, speaks of risk control, while DIN EN ISO 3100 uses the term risk management. However, risk treatment or risk control are also possible.[157]

What is common to all, however, is that it is a matter of deciding how the company will deal with the risks that are now known. The aim is to determine measures that are acceptable and feasible for everyone and have a high benefit, but do not burden the costs too much.

As already explained, risks are assessed on the basis of their probability of occurrence and the amount of the expected loss. This results in a company-specific risk situation. Appropriate measures are therefore aimed to

- Reduce the probability of occurrence,
- Limit the impact,
- Change the risk situation,
- Strike a balance between opportunities and threats.[158]

According to the COSO framework, risk management involves a range of tools and measures to align risks with the company's strategically defined risk appetite and tolerance.[159]

The international standard DIN EN ISO 31000, on the other hand, speaks of risk management, which includes the selection and implementation of various suitable measures for dealing with risks in terms of quality objectives and tolerability within the framework of the risk policy.[160]

This phase in the risk management process is aimed at influencing the company's risk situation and achieving a balance between opportunities and threats in order to increase the company's value and achieve its corporate goals.

Appropriate measures can avoid intolerable risks and tolerably reduce high unavoidable risks.[161]

---

[156] Hoffmann (2017, pp. 22–25).

[157] Hoffmann (2017, p. 43).

[158] Hoffmann (2017, p. 43).

[159] COSO – The Committee of Sponsoring Organizations of the Treadway Commission (2004, p. 4).

[160] DIN ISO 31000, p. 28.

[161] Romeike (2018, p. 44).

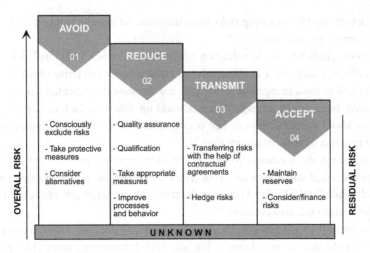

Fig. 1.25 Variants of measures in dealing with risks

Risk management is the responsibility of the risk owner (manager) and is carried out in compliance with the specified framework conditions of the company-specific risk policy.

According to DIN EN ISO 31000, risk management as part of risk management also has a process-oriented structure. The standard identifies the following steps:

1. Assessment of risk management,
2. Determination of the tolerability of the residual risks,
3. Elaboration of the measures,
4. Assessment of effectiveness through monitoring.[162]

The selection of the coping strategy should also take into account any secondary risks that have not yet been considered in the management of the risks already identified. Monitoring and control of the measures used should therefore constantly reassess and adjust the respective strategy.[163]

Measures to control or manage risks can be active – cause-oriented – or passive – effect-oriented. In the case of active risk management, an attempt is made to eliminate or reduce the corresponding causes (cf. Fig. 1.25).

Active measures lead to the avoidance or reduction of the probability of occurrence or the amount of damage. Passive risk management, in contrast, does not change the risk

---

[162] DIN ISO 31000, p. 28.
[163] ISACA Germany Chapter (2014, p. 20).

structures, but accepts the risk and attempts to reduce the impact. Passive measures are risk transfer or the acceptance of risks without adjustment.[164,165]

The goal of risk avoidance is to avoid taking a risk in the first place. This can be done by choosing alternative options or by completely foregoing the risk. Risk avoidance makes sense in the case of particularly high risks or when the accumulation of multiple individual risks means an incalculable situation.[166]

Risk reduction usually only occurs when risks have already occurred and active action must be taken in a timely manner. The aim is to reduce the risk parameters to a residual risk that can be accepted by the management and the stakeholders.[167]

In the case of risk transfer, the overall risk is passed on to a third party, for example to an insurance company, or divided among various parties involved, for example through contract design among subcontractors and suppliers. This mitigates and reduces the potential impact on the company.[168]

Risk acceptance means that risks are simply accepted. This happens above all in the case of residual risks with only a low probability of occurrence and impact or if the costs of risk treatment would be higher than the expected damage caused by the risk. What is important here is that applicable rules and regulations are adhered to and neglected for the sake of preserving opportunities.[169]

DIN EN ISO 31000 additionally shows as an alternative measure the taking or even increasing of risk in order to take advantage of an opportunity. It is important that this measure does not lead to non-compliance with or circumvention of existing laws and regulations in order to achieve an advantage.[170]

In order to select the right measures, the identified risks must first be prioritized on the basis of defined criteria, as it is usually not possible for the company to address all risks. In doing so, the resulting costs or disadvantages of the risk, possible opportunities and the costs for avoiding or reducing the risk are weighed up against each other and thus a list of priorities is determined.[171]

When selecting the appropriate strategy and measures, it is important to weigh the costs and benefits of risk management against the economic interests of the company and legal or social requirements.[172]

The defined measures are collected in a plan. This clearly documents the reasons for choosing each measure for each risk, where the responsibilities lie, which activities are

---

[164] Hoffmann (2017, p. 43).

[165] Romeike (2018, pp. 44–45).

[166] Hoffmann (2017, p. 44).

[167] Hoffmann (2017, p. 45).

[168] Hoffmann (2017, p. 45).

[169] Hoffmann (2017, p. 47).

[170] DIN ISO 31000, p. 28.

[171] Brüggemann and Bremer (2020, pp. 51–52).

[172] DIN ISO 31000, pp. 28–29.

recommended, what resources and time are required, and how the results will be measured and reported.[173]

Before, during and after the implementation of each individual measure, it should be checked for effectiveness and efficiency and improved if necessary.[174] This monitoring forms an important part in the overall risk management process. Controls can be carried out on a regular basis or when the need arises.

According to DIN EN ISO 31000, the monitoring and review in the risk management process leads to the following effects:

- Ensuring effective and efficient risk control,
- Improving risk assessment,
- Learning effect from experiences, changes, further developments, successes or failures,
- Clarification of changes within and outside the company that result in an auditing of the risk policy,
- Identifying new risks.[175]

The risk management process presents itself as a cycle. By monitoring and controlling the measures, opportunities for improvement can be recognized and new risks identified. These are analyzed, evaluated and prioritized again. Opportunities and threats are weighed against each other and new measures for risk management are defined.

1.3   Process of Building a Structure

A project is generally characterised by the following features

- It is unique,
- It has a defined goal,
- Ist has a defined beginning,
- It has a defined end.

This is determined by the elements of costs, deadlines, qualities, quantities and, in addition, by information, contracts and insurances as well as overarching organisation, coordination and documentation.

Construction projects are characterized by three essential fields of activity: project management, planning and execution. Before the actual execution of the construction work, and thus the creation of the building, extensive preliminary considerations are

[173] ISACA Germany Chapter (2014, p. 21).
[174] ISACA Germany Chapter (2014, p. 20).
[175] DIN ISO 31000, p. 30.

necessary, which must be made within the framework of the planning process. Project management is installed to ensure that planning and execution are implemented in the necessary quality, within the estimated time and at the planned cost.

1.3.1 Project Management

Especially in construction projects, the individual character of each project is essential. Compared to process management, predefined processes and methods can only be integrated to a limited extent here, as each construction project brings with it its own special features and needs.

Thus, projects according to DIN 69901 are basically characterized by the fact that they have both a defined beginning and a defined end. Furthermore, they have individual characteristics. Consequently, they are to be understood as unique. Another characteristic of projects is the limited availability of resources. These characteristics apply to construction projects in particular.

Countless work steps are required for the construction of a building. These influence each other to a great extent. Moreover, they do not take place under controlled conditions of a stationary production. Rather, construction sites are subject to external influences that are often beyond the control of those involved. These include, for example, influences from the building ground or the weather. Thus, the influencing work steps cannot all be calculated exactly in advance. The construction project thus becomes a complex project. The project is handled by project management. According to Koelle, project management is

> primarily the art of getting the desired work done by people within the promised time and available resources with success.

Project management pursues the goal of defining, initializing, controlling and finally successfully completing a project with the help of organizational structures, methods, techniques and resources.[176]

Project management is based on a specific service profile. In addition to the services provided by architects and engineers, the service profile of the project manager has constantly evolved in line with the requirements of today's construction projects. Derived from the services of project control, project management was established as an independent professional title.

In addition to the project team on the part of the client, project managers as client representatives with authority to issue instructions or project controllers in an advisory capacity can form the upper management level. If the project manager is active in a staff function as an advisor to the owner, project management is an independent area of responsibility in the handling of construction projects.

---

[176] Madauss (2017).

Especially in the last decades, the demands on a building or a building complex have increased considerably. With these increasing requirements, the complexity of the work steps to be implemented has also increased. Due to this complexity, the demand for project control services has increased. A project controller is particularly active in the areas of organisation, costs, deadlines and quality. His essential core tasks include project controlling and reporting to the client.

From this advisory function, project management has developed into an essential service in the construction industry. Project management is to be differentiated from project control in that the project manager assumes the client functions assigned to him and is thus authorized to issue instructions. The project controller, on the other hand, ensures that tasks are performed and decisions are made, but does not do this himself (cf. Fig. 1.26).

In the construction industry, the remuneration of planning and consulting services is regulated by the Fee Structure for Architects and Engineers (HOAI). In addition to the architects and engineers of various specializations, the project controller or the project manager belong to the circle of professional participants, especially in complex construction projects. Both act on the management level.

The "Committee of the Associations and Chambers of Engineers and Architects for the Fee Regulations e. V." (abbreviated to AHO) set itself the task of delimiting and differentiating the service profiles of the project controller and the project manager from the service profiles of the architects and engineers (cf. Fig. 1.27). The aim was to avoid an overlapping of services and subsequently to ensure a basis for the remuneration of project control and project management services.

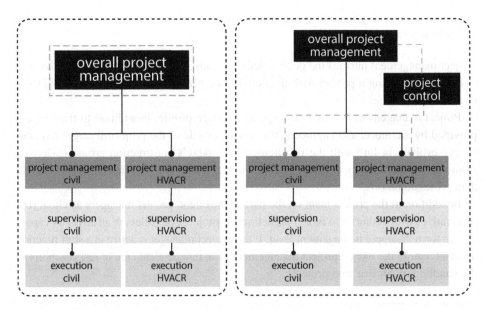

Fig. 1.26 Project control and project management services

service phases (HOAI)

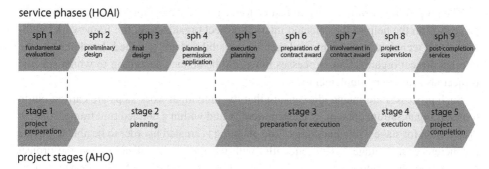

project stages (AHO)

Fig. 1.27 Comparison of AHO and HOAI

Project management tasks are assumed by the project manager as the representative of the client. Due to this development, the AHO decided that the fee basis developed for this purpose, together with the service profile, should be understood "as the sum of project control and project management tasks".

Consequently, the services of project management are closely linked to the service profile of project control. Project development, management of the users of a property, risk management and other service areas offer further catalogues of project management activities.

The AHO describes the services of project management according to basic services and special services, based on the regulations of the HOAI. Whereas the HOAI divides the services of architects and engineers into nine service phases, the AHO structures a project into five project stages (cf. Fig. 1.27).

Project Stage 1: Project Preparation
Stage 1 of project preparation essentially involves project initialisation. Here, the definition of project goals and the updating of the goals are of particular importance. Furthermore, various analyses and evaluations are to be prepared in order to determine the greatest possible benefit of a construction measure to be implemented. These include demand and feasibility studies, space and room programs, studies of possible locations as well as investment and return on investment considerations.

In order to make the decision to implement a construction project, the expected costs must be compared with the benefits. In principle, the general conditions for the implementation of a construction measure are analyzed and evaluated in the project initialization. Based on these findings, the project objectives are updated.

The project initialization is accompanied by the project organization. This provides the framework for the project. The project organization structures the communication between the participants, regulates the decision-making, change and risk management, and defines the initial quality, cost and schedule frameworks.

The project organization must do justice to the respective complexity of a construction project. With the help of the selected internal project organization, the framework conditions of all technical participants and their responsibilities are specified. In addition to the internal project organization, processes must also be defined according to which specific project tasks are to be implemented.

Such processes are intended to ensure that the individual work steps are carried out in terms of organisation, content, consistent quality and within a defined timeframe. Also, in preparation for stage 2, the planning of the planning is created in order to be able to pursue a concrete goal in stage 2. Consequently, with the project initialization and the project organization, the foundations are laid for the implementation of the further project stages.

In summary, the main tasks of project management are in stage 1 – project preparation:

- Analysis of suitable sites or uses for a site,
- Feasibility Analyses,
- Demand planning,
- Functional and area programs,
- Definition of project goals,
- Communication (correspondence, reporting and meetings),
- Decision, change and risk management,
- Cost structures – and cost optimization,
- Financial Planning,
- Time Frame,
- Planning the Planning,
- Contract Management,
- Procurement and contract structures,
- Catalogues of services Fees,
- Insurance Concept,
- Logistics.

In all these analyses, the risk analysis in the project initialization is to be emphasized in particular. Based on the analyses for the project, the project goals are defined. All further considerations regarding costs, deadlines and quality are based on these project goals.

If, in the further course of the project, findings are made that lead, for example, to a change in the underlying demand planning, this can have global effects on the construction measure. The change of a demand would possibly also be accompanied by a change of the space programme. Such a change could entail new analyses, evaluations, planning and effects on costs and deadlines.

This extent is due to the fact that all the considerations and specifications for initialization are interlinked. Due to this interconnection, a culture of communication that is not practiced in the project, but also in companies, harbours considerable risks.

Project Stage 2: Planning

With the help of the planning of a project, it should be described which requirements are to be fulfilled in order to achieve the project goals. Here, the definition of project goals and the updating of the goals are of particular importance. The group of planners includes different disciplines depending on the complexity of a building project. Based on the requirements planning, the architect often creates or deepens the space and area program. Internal processes for the use of the building are also examined here and used as a basis for planning.

In addition to this functional planning, it is also the architect's responsibility to initialise the design of the rooms and the surroundings. This analytical and at the same time creative examination of the building project is shaped to a great extent by the demands of the later users as well as the functional and technical contexts.

In addition to the architectural planning, the technical planning is carried out with regard to the technical requirements and qualities. These considerations include the supply of the building with media such as heat, water and electricity, but also safety-relevant fire protection requirements. Special technical planning such as acoustic planning for concert halls, planning for laboratories or the design of high-security areas can also have an influence on the technical building equipment in the same way as on the architectural planning.

In this planning phase, the exchange of information and communication between the decision-makers and planners is particularly important. Originally, the coordination of the construction project was the task of the coordinating architect. Particularly in the case of larger or complex construction projects, a project controller or project manager is often called in addition to the coordinating architect. This project manager acts as the client's representative.

In stage 2 planning, the focus is on the planning with all the considerations made and the links between the various specialist disciplines. In addition to the planning concept itself, the decision and change management as well as in-depth cost and schedule considerations must also be carried out with the approval procedure.

In project planning, i.e. the stage opposite the preliminary design and design planning phases, the focus is on cost estimation, cost calculation and life cycle cost calculation with regard to risk assessment. These serve as an initial estimate of the expected investments in order to compare them with an expected return. The life cycle calculation in particular can take into account various scenarios within the service life of a building.

The objective of project planning is to describe all requirements for components and building elements, products and equipment, and to ensure that the planned work is fully functional upon completion. Level 2 project planning includes in particular:

- Planning concepts,
- Information Exchange,
- Approval Process,
- Implement decision, change and risk management,
- Review, analyze and evaluate planning results,

- Analysis and evaluation of reference objects and product sampling,
- Cost optimization of the planning phase,
- Cost control and cost optimization,
- Estimating occupancy costs and building operations,
- Scheduling of the call for tenders and awarding of contracts,
- Deadline update within the planning phase,
- Rough logistics schedule,
- Contract Controlling Planner.

Once all relevant project objectives have been defined and planned, the execution preparation phase can begin. In project stage 2, the group of people involved in the project has expanded.

Against this background, the performance of the planners must always be checked with the project management within the contract controlling. It is also the responsibility of the project manager to always compare all planning results with the project goals and specifications. Due to this complexity, project managers from different disciplines form a team in order to be able to communicate with all specialist planners on an equal footing.

This project stage also conceals various risks. In particular, the costs and deadlines to be established are not subject to any concrete parameters here, but are based on estimates. Incomplete planning or failure to update the plans can also have an impact on the upcoming awarding of contracts for services. If the requirements for the services to be commissioned are not clearly described with the planning and construction description, then not only effects on costs and deadlines, but also on interfaces to this service can be expected subsequently.

Project Stage 3: Preparation for Execution
The stage of preparation for execution comprises the concretisation of the design planning into an execution planning as well as the awarding of construction contracts in order to bind companies for the execution of the construction measure.

In addition to the planning specifications, the individual services are described comprehensively and adequately within the framework of a performance specification. This description must be sufficient to enable construction companies to convert the tendered services into an offer.

This description includes the construction description, the planning and service specifications, which tender specific services and products. As a rule, invitations to tender for construction services are published on an online tendering platform, so that any company operating on the market that is capable of providing these services can submit a bid.

The cost of construction services, like many other products and services, is highly dependent on the state of the economy. In times of housing shortages, a backlog of investment in infrastructure buildings and a far-reaching restructuring of the energy supply, demand is higher than supply – prices rise as a result.

When a construction work is put out to tender, the day of the submission is also announced. On this day, all offers are opened and subsequently analysed in order to finally award the construction contract to one of the bidders. In particular, the bid prices and the references of a company are checked.

Unit price contracts are often concluded in the construction industry. Since the tendered scope of services has been prepared on the basis of an implementation plan (sometimes also a design plan), which is subject to inaccuracies, the tendered quantity of a service may deviate from the quantity actually provided. The quantity actually performed is then invoiced at the agreed unit price.

If further services are required in the further course of a construction project, changes in the execution of the building are demanded by the client, supplementary orders can be made on the basis of the agreed unit prices. So-called supplements (to the main contract) are to be agreed for such contractual supplements, which often only become apparent in stage 4.

In addition to the construction content, i.e. the construction work itself, the construction circumstances, above all the expected construction period, are also decisive for pricing. The construction period has a major influence on the costs of execution, particularly from the point of view of construction operations and construction economics. Thus, on the basis of the performance specifications, a forecast is made for the duration of the execution of these partial services and, in the context of the overall construction measure, brought into a sensible sequence as well as in connection with further services.

How long it takes to provide the agreed services depends in particular on the personnel and equipment resources deployed. These plans can also involve numerous risks. Consequently, risk considerations by the project manager must also be taken into account in project stage 3.

The responsibilities of the Project Manager at Level 3 include:

- Supervise and monitor the exchange of information,
- Initiate and organize project meetings as needed,
- Implement decision, change and risk management,
- Quality Control Planner,
- Review procurement procedures,
- Organize and implement reporting,
- Review and participate in bid evaluation,
- Cost control,
- Scheduling,
- Prepare and negotiate contracts for contractors,
- Management of security services by the construction companies,
- Prepare supplement management,
- Contract controlling including contractually agreed dates and deadlines.

Project Stage 4: Execution
The project stage 4 execution is usually the longest and most extensive project stage in project management.

A selection of services according to the service specifications drawn up by the AHO are:

- Update organizational structures,
- Grant and monitor information flows,
- Organize and implement reporting,
- Control of the technical participants,
- Quality controls for fee and construction services,
- Acceptance and functional tests of services provided,
- Defect Tracking,
- Cost control and optimization,
- Scheduling,
- Contract and Subcontract Management,
- Checking invoices.

Within the stage of execution, the now commissioned construction work is carried out. Now the actual performance on site begins. This project stage, which is equivalent to service phase 8 according to HOAI, is often the most intensive phase in a construction project.

Ideally, all qualities and objectives have been defined and commissioned in the planning and performance specifications at the start of this phase. These descriptions are described as construction target or target process in terms of scheduling.

With the execution of the construction work, the examination of the actual situation with regard to the progress of the construction work, the deadlines and the costs begins. The responsible (specialist) planners monitor the construction work of the trades or construction companies assigned to them.

In addition, the responsible (specialist) planner is also entrusted with the quality control of the executing construction companies. With the continuous observation of costs, deadlines and quality, reports often have to be submitted to the building owner at regular intervals. The content of these observations is the comparison of the planned target situation with the current actual state.

With regard to the complexity of this task, the external factors of a construction measure must also be taken into account. In Germany, for example, it is common for the phases of planning, preparation for execution and execution to overlap. From the client's point of view, this has the advantage that processes and tasks can be processed in parallel if project phases overlap.

However, it is particularly evident in the case of complex construction tasks that the probability of risks occurring increases due to the overlapping of these phases. In order to keep an eye on the development of the actual construction process in comparison to the planned course, a common instrument of project management is the "target/actual comparison". This is usually done on the basis of the "magic triangle" of project management. This looks at:

* Cost,
* Dates,
* Qualities or performance.

At regular intervals, usually every week or month, the actual situation is compared with the originally planned target situation. By evaluating and analyzing the actual situation against the planned progress, knowledge can be gained for risk management and further risk avoidance. This is one of the core tasks of the project manager in project stage 4 execution.

Trend analyses represent a further instrument. These trend analyses bundle the project-specific information and map it in the context of an expected performance progress – i.e. a forecast. With the help of the trend analysis, conclusions can also be drawn with regard to the adherence to interim targets and the execution speeds.

Basically, stage 4 execution involves various risk scenarios, since a lot of new information is brought to light during construction. Here, it is crucial that project management achieves continuous and consistent control of information and decision-making.

Project Stage 5: Project Completion
In the context of project closure, various measures are initiated to dissolve the project organization. The main tasks of stage 5 project closure consist of evaluating and analyzing the construction measure that has now been carried out. In particular, the focus is on comparing the originally planned target performance with the actual performance that has now been achieved.

The project management must comprehensively document the extent to which changes or deviations occurred in the project. It must also be ensured that all significant decisions, problem solutions and coordination are documented with the help of the project documentation. The aim is to use the documentation to fully present the course of the project and make it comprehensible.

Furthermore, all supporting documents confirming the formal requirements for a construction project must be compiled in full in the construction documentation. This documentation is a comprehensive compilation of all products and systems installed in the building project. All calculations, such as the statics or calculations of the building services, must also be included in the construction documentation. For complex systems in the building, such as building automation, instructions for the operation of systems must also be included so that the later user is able to "operate" the building.

Before the building project is handed over to the client or the user, various test phases and adjustments have to be carried out with regard to the building technology. For example, after cleaning the construction site, the ventilation systems are put into trial operation and the air flows are adjusted. Test runs are also carried out with regard to the measurement, control and regulation technology. Once the construction project has been formally handed over to the client, the project organization is dissolved.

Stage 5 is essentially about collecting and handing over all project-specific information. With the help of the documentation of all relevant decisions, changes, problems, and solutions as well as the executed services, the construction of the building is to be documented in a verifiable manner.

In addition, further target/actual comparisons are made, such as the comparison of the cost estimate with the final cost statement. Based on the evaluation of the cost acuity and the deviations between the cost estimate and the cost statement, conclusions can be drawn about the quality of the cost planning.

A further step in project completion is to document and evaluate the empirical values. Thus, supposedly identical construction tasks can entail the most diverse project structures and thus differentiated requirements for project implementation.

Based on this knowledge, it is possible to identify and minimize potential risks in other projects. This is particularly relevant against the background that construction projects represent individual projects.

The project stage 5 completion provides for the following exemplary services for the scope of services according to AHO:

- Documentation,
- Acceptance and rectification of defects,
- Commissioning,
- Cost Determination,
- Audit,
- Site logistics,
- Contract Controlling,
- Management Security Retentions.

In addition to the project stages, the services of project management have been subdivided into five areas of action by the AHO since 2004. These five areas of action are:

- Action Area A: Organisation, information, coordination and documentation,
- Action Area B: Qualities and quantities,
- Action Area C: Costs and Financing,
- Action Area D: Deadlines, capacities and logistics,
- Action Area E: Contracts and insurance.

The five areas of action are reflected in the basic services of each of the five project stages.

The Action Area **A** "Organisation, information, coordination and documentation" is to be understood as a superordinate field of action to the other fields of action.

It is the responsibility of the project manager to define the project-specific organisational structures and to document these definitions and agreements.

This includes, in particular, communication management, which also includes information and reporting. Decision-making, change management and risk management are also assigned to area of action A.

Action Area **B** comprises the controlling of qualities and quantities with regard to the requirements planning and requirements for the structure to be built on which the project initialisation is based. If changes occur, the objectives of the construction project must also be updated with regard to qualities and quantities.

Based on the organisational structures and qualities to be defined, the costs and possible financing are to be set up and considered within the scope of Action Area **C**. The costs and possible financing are to be set up and considered within the scope of Action Area **C**. The costs and possible financing are to be set up and considered. Here it is not only essential to create a cost framework, but also to repeatedly subject this to economic consideration with the project-specific cost tracking.

In a next step, the deadlines are to be coordinated at different levels of detail in the Action Area **D**. Logistical interrelationships on the construction site also play a role here.

In Action Area **E**, corresponding contracts and, if necessary, required insurances are to be concluded. This area of activity essentially includes the design of award procedures for construction and planning services. Involvement in the selection of bidders and the conduct of negotiations are also anchored in area of action E.

Furthermore, the so-called special services are taken into account. The special services are to be understood as optional services of the project management. These services should serve to meet the individual construction task. The essential services of the project manager include the participation, the coordination, the examination, the analysis and evaluation of facts. This is always done against the background of achieving the project goals.

1.3.2   Planning Services

The idea comes before the deed. This general consideration applies in particular to building projects. The idea of the building to be erected must be advanced and written down to such an extent that a third party can understand it in the same way. Planning is considered to be the architect's very own task.

During the planning phase, the architect must skilfully arrange the functional areas in order to ensure that the building is as useful as possible. Once the spatial program has been harmonized with the other functional contexts, the appearance of the building is to be planned on the basis of these parameters. In conventional planning, the architect prepares a design plan on the basis of the needs analyses prepared in advance and other conditions placed on the building.

For the building application to be submitted, the agreed draft planning is detailed up to the planning maturity of an approval planning. The approval planning, which must primarily meet formal requirements, is then prepared for the building application. Once the building application has been approved by the responsible municipality or the state, the next step is the preparation of the significantly more detailed implementation planning.

In the course of the elaboration of the implementation planning, the other technical parties involved are included. Modern buildings are subject to various requirements with regard to their energy balance and the material cycle of the building materials used. In addition, the requirements for the operation of a building and the safety requirements have increased significantly in recent years.

As modern buildings now house many automated processes in operation, the evolution of modern buildings can be interpreted as a move towards a machine. Due to this multitude of influences, the architect must be assisted by specialist planners.

The most essential specialist planning services comprise the planning of the interrelationships of the building services engineering and building air conditioning, the electronic equipment and components as well as specialist services relating to the special features of the building to be erected. Under the leadership of the coordinating architect, all information relevant to the building is thus coordinated and compiled in the execution planning during the planning phases. An essential task in the compilation of the individual plans is the so-called collision planning.

In the context of collision planning, the various hand-drawn plans were originally superimposed as transparent ones in order to identify possible overlaps of components and collisions in the design of the construction elements and to revise them.

With the introduction of CAD programs (Computer Aided Design), the various technical plans could be created with the support of computers. As a result, the plans could be digitally superimposed and, in a further step, examined for collisions via the elaboration of 3-D models from different viewing levels. In addition to the comparison of the plans, the representation of three-dimensional buildings had a particularly strong influence on the architectural competition, since the dimensioning of a design created a new medium for the representation of the building's appearance.

Such planning maturity is not only characterized by the fact that the structure to be executed is drawn with a high degree of detail, but rather that what is depicted or described should function as a finished work in the end. This requires that the numerous functionalities of a structure are planned and calculated correctly.

This starts with basic requirements for the stability and serviceability of the structure, through more far-reaching requirements for economic and environmentally friendly operation, to aesthetic requirements for the design. It is common for planning to take place in parallel with execution (cf. Fig. 1.28).

In a next development step within the construction industry, the method of Building Information Modeling (BIM for short) was introduced. The basis for this methodology is the three-dimensional representation of the building design. With the help of the BIM methodology, information is assigned to the 3-D model of the building at various levels. For example, a window is not only represented in its appearance in the 3-D model, but all other information about this window is also stored.

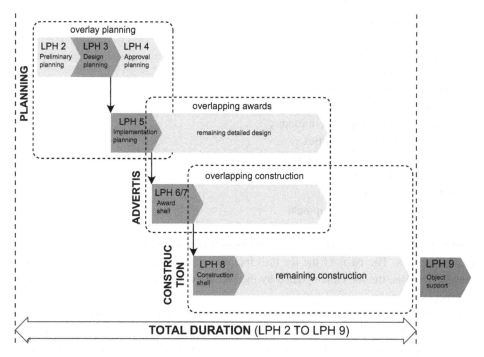

Fig. 1.28 Time overlap of the service phases

In this way, the required number of the specific window type can be determined, the prices stored and the energetic properties of the window itself assigned. Within the framework of the various service phases of the architect or the project stages of the project manager, concrete information can be compiled, for example, for the creation of a bill of quantities as part of the award process.

Changes can also be implemented in the 3-D model during the construction process, continuously checked for possible collisions and supplemented with their respective information. This is particularly enriching in the case of parallel service phases. If this methodology is implemented, the documentation of each change as well as the digital representation of the final building is automatically carried out. Consequently, information is also prevented from being lost in the planning and execution process. This 3-D model can then be updated throughout the entire life cycle of a building.

The remuneration and the service profiles of planning are legally regulated by the Fee Structure for Architects and Engineers (HOAI). With its currently valid seventh amendment from 2013, the HOAI takes into account a total of 14 different service profiles in the structuring of the remuneration for such services.

These performance patterns are:

1. Consulting Services,
2. Land Use Planning,

3. Development Plan,
4. Landscaping,
5. Green space planning,
6. Landscape framework planning,
7. landscape conservation planning,
8. Maintenance and development planning,
9. Planning of buildings and in rooms,
10. Planning of outdoor facilities,
11. Design of engineering structures,
12. Planning of traffic facilities,
13. Structural Design,
14. Planning of technical equipment.

Within the respective service profile, a distinction is made between basic services and special services. The basis of the fee thus includes the service profile, the fee zone, the chargeable costs, the fee table updated by the HOAI as well as other possible surcharges and special services.

Once the scope of services has been determined, the fee zone is determined in a second step. The fee zone serves to classify the expected degree of difficulty associated with the construction task. Thus, fee zone I comprises the assessment of the services as very low planning requirements. The highest fee zone V is to be used for very high planning requirements. The fee zones were created in relation to the scope of services.

Allowable costs are concrete cost parameters from the cost structure that reflect the construction volume to be implemented in terms of costs. Thus, the chargeable costs for the construction of a single-family house are lower than the costs for the construction of an airport. Nevertheless, the fee zones, which reflect the degree of difficulty, can be the same.

Experience shows that there are high planning requirements for the realisation of an airport, as this construction task is subject to a wide variety of aspects. A single-family house, on the other hand, accommodates only one family, but the requirements for the single-family house to be built can nevertheless be very high. For example, it would be possible that this single-family house would have to meet very high energy requirements. Furthermore, it would be possible that the family wants a special room layout and functional processes within the single-family home. Another possibility to prove a high requirement profile would be a comprehensive building automation.

Once the fee zone and the chargeable costs have been determined, the corresponding fee tables from the Fee Regulations for Architects and Engineers are to be used as a basis. This fee schedule is intended to ensure that the remuneration of planning services is subject to a uniform and legally regulated framework.

Furthermore, in the case of conversion and modernisation measures, corresponding fee supplements are taken into account in order to cover the increased expenditure of an existing building and the risks to be expected with it in terms of fees.

The success of the work is essential for the remuneration of the services described in the HOAI. The architect, like all specialist planners, owes a work performance which includes the defect-free construction of the building.

However, it is not only architects and specialist planners who draw up plans. Thus, it is incumbent on the specialist companies, i.e. the executing construction companies, to create an even more detailed so-called work and assembly planning. The work and assembly planning does not represent a change to the execution planning, but a concretization.

The various plans therefore form the foundation stone for the construction work to be carried out on site. On the basis of the planning documents, all coordination of design-related technical and functional importance takes place.

1.3.3   Construction Work

The tendering, awarding and invoicing of construction work is regulated by the German Construction Contract Procedures (Vergabe- und Vertragsordnung für Bauleistungen, VOB). The VOB is composed of:

- VOB Part A – General provisions relating to the award of construction constracts,
- VOB Part B – General conditions of contract relating to the execution of construction work
- VOB Part C – General technical specifications in construction contracts

If the VOB is the subject of the contract, this also includes the VOB Part C: General technical specifications in construction contracts. The VOB Part C currently consists of 64 DIN standards (ATV DIN 18300 – "Earthworks" to DIN 18459 – "Demolition and Dismantling Works"), which contain trade-specific information on the execution of construction work, among other things. Various service areas or trades are, for example:

- Demolition and deconstruction work,
- Concrete work,
- Roofing work,
- Insulation and fire protection work on technical installations,
- Drainage and infiltration works,
- Electrical, security and information technology systems,
- Earthworks,
- Screed work,
- Tile and slab work,
- Track work,
- Explosive ordnance disposal,
- Plumbing,
- Bricklaying,

- Metalwork,
- Plaster and stucco work,
- Reinforced concrete work,
- Carpentry,
- Shoring,
- Curtain-type ventilated facades,
- Carpentry and woodwork.

The current version of the Standard Service Book for Construction (STLB-Bau) published by the Joint Committee on Electronics in Construction (GAEB) and the Main Committee of the German Contract Awards and Contract Committee for Construction Work (DVA) actually covers 77 trades. These start with works for the construction site equipment, lead to works of civil engineering, structural engineering, non-technical and technical finishing works up to weather protection measures.

With very few exceptions, construction work is not carried out by a single person. A large number of people are involved in completing a building to the point of functional maturity. The many subtasks that have to be performed in the process are also referred to as trades.

An individual company can be contracted for the award of construction services for a specific trade-specific service. It is also possible to award a construction management contract as a package of several contracts to one company.

Furthermore, it is possible to involve a general contractor. This is then responsible for the execution of all construction work. A general contractor usually combines large construction companies in order to be able to offer construction services from as many different trade groups as possible.

It is also common practice in the construction industry for so-called consortiums – ARGE for short – to form in order to bundle their range of services for a specific project and to jointly implement the services put out to tender. In the case of a large scope of services to be provided by one company, other companies are often bound as so-called subcontractors. Particularly in the case of large construction projects, construction companies are unable to muster the necessary resources on their own and thus bind further construction companies within their contractual performance.

Planning of the Construction Audit

2

2.1 Introduction

Construction auditing is often not perceived by the departments subject to auditing as a helpful tool for improving their own work. Rather, the audit must be prepared for its work to be perceived as disruptive, annoying or even counterproductive.

It is therefore to be expected that information and documents will not be immediate, complete and self-explanatory. This situation must already be taken into account in the preparation of the construction auditing. Following the principal-agent theory, this can be summarized as shown in Table 2.1.

A well-functioning audit planning ensures that the resources of the construction auditing are used efficiently and effectively exactly where the company is exposed to great risks in the project execution on the one hand, but can also possess great opportunities on the other hand.[1]

Experience has shown that construction audits offer little benefit if they are not carefully prepared. For example, it is often found that projects are audited in areas that show little or no reason for fraudulent acts, or are too small or not complex enough.

For example, it makes little sense to review the change management processes if the test project is, for example, the replacement of flooring in a small administrative building, this service is also the only construction service, there is no time pressure, no planning changes are expected and the construction service is exhaustively described.

For such a project, on the other hand, the review of the award of the construction work would be more interesting due to the small size of the project, as all services were very likely awarded in the sub-threshold range. This means that the increased requirements for

---

[1] Eulerich (2018, pp. 224–225).

Table 2.1 Challenges of the reviews

Aspect	Hidden information	Hidden actions	Hidden properties
Cause	Information imbalance, unobservable level of information	Unobservable activities, unrecognizable agreements in the background, insufficient willingness to cooperate	Behaviour and working methods not known in advance, as well as performance and judgement skills
Effect	Results based on incomplete data, lead to wrong conclusions	Gaps in documentation are closed retrospectively, deficits are concealed, evaluation of results is made more difficult	Implementation of the audit more difficult, evaluation of alleged deficits prone to error
Solution	Providing information about one's own work, communicating work steps, installing incentive and control systems	Intensive monitoring of the auditee, install incentive and control systems	Intensive observation of the person to be examined, implementation of cross-checks in the interviews

public tendering and awarding above the thresholds do not apply, which in turn creates room for fraudulent acts.

On the other hand, it is observed that for very complex and large projects with contract sums above the thresholds, the tendering and contracting of works is checked for individual awards, although a check of the designer's performance during service phase 8 (construction) would be more relevant, since the supervision and management of a construction project with many individual trades requires a structured and systematic way of working.

These services of the planner are in turn not clearly regulated by any legislation or standard. It is true that the Fee Structure for Architects and Engineers (HOAI) contains information on the scope of services in the respective execution phase. However, experience has shown that these are filled out differently by the planners commissioned in each case, since in the opinion of the authors, the service specifications in the HOAI lack detail. Such a level of detail can only be created individually through agreements with the planner supervising the construction.

These examples show that the selection of the areas or processes to be audited, and thus the direction in which the preparation of a construction audit will go, will depend very much on the size and complexity of the project.

Another significant factor is the type of contract. For example, a project that has been handed over to a general contractor will not require a review of the award of construction contracts, as this is a service that has been transferred to the general contractor and which the latter performs on its own responsibility.

In the case of such a contract, on the other hand, it will be more interesting to check before awarding the contract whether the tender documents for the general contractor cover all the risks of the project or whether the tender documents contain various gaps which would have to be "plugged" later by means of expensive amendment agreements.

This contrasts with projects whose service areas are completely divided into individual contracts. For such projects, the review of project management and contract management is of high importance to ensure that the interfaces between the individual contractors do not have any gaps but also no duplications and that the client does not lose the overview of the project.

Similarly, it will be important for the preparation of a construction audit whether it is a new construction project or an existing project. Let's take the example of the new construction of a school and the modernization of one. The goal is the same for both projects. The end goal is to have a finished and usable school in time for the start of school. However, the challenges to the execution of these projects are unequal, as there are different risks in the execution.

In the case of the new school building, unforeseen problems in the subsoil could lead to delays despite the subsoil expertise. This could be due to the discovery of munitions or unexpected contamination. The new building itself should proceed without any problems, especially if it is possibly built in modular construction and the design has already proven itself at other locations or is not being built by the executing company for the first time. Risks from changes to the construction design are therefore hardly to be expected.

The modernisation of the existing building, on the other hand, could cause surprises if components unexpectedly prove to be no longer sufficiently load-bearing during the conversion measures and may have to be replaced. Changes are therefore more likely to be made to the existing building. Risks from the building ground, on the other hand, are not to be expected.

In this respect, the weighting in the selection of the test fields will be very different for the two schools.

However, another important factor influencing the planning and preparation of the construction audit is the timing of the audit. Thus, a project-accompanying audit (ex-ante consideration) requires different preparation than a one-time and retrospective review (ex-post consideration). And the preparation for an event-related review of a selected process (ad-hoc consideration) will again require different preparation.

And last but not least, it will be important for the preparation of a construction audit which objective is pursued with the audit. The selection of suitable audit areas and a suitable project will be different if an owner merely wants to preventively review and improve his processes in the execution of a construction project or if this owner is specifically looking for fraudulent acts or cost savings. Figure 2.1 shows the most important influences on the planning of a construction audit.

This chapter will therefore focus on identifying the right objectives, audit areas, projects and scheduling. To this end, tools and suggestions will be presented to help you sharpen your own audit planning in detail.

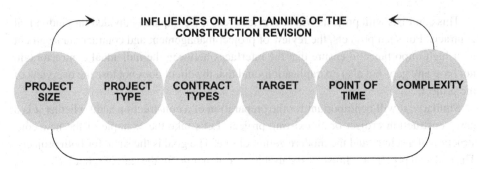

Fig. 2.1 Influences on the planning of the construction audit

2.2 Audit Objective

Careful preparation and planning of a construction audit is the prerequisite for finding and achieving the right objectives for the construction audit. This is where the first challenge for the initiator of a construction audit becomes apparent. For the success of the measure, it is necessary that the initiator knows exactly what the objective is and has precise expectations of the result.

So as the initiator of a construction audit, you should first ask yourself what you want to achieve. Are you concerned that your construction projects are not properly managed and organized? Do you believe that because of this, your processes are ineffective, causing your employees to lose a lot of time and become overwhelmed? Or do you even believe that your processes are not sufficiently defined and are therefore lived differently by your employees, resulting in a loss of quality? Perhaps you also fear that your processes are even incomplete and urgently need to be supplemented.

Or do you think you are losing money because construction services are not being billed properly or you are paying for more design services than you actually receive. You may also have both fears. Risner therefore quite deliberately makes a fundamental distinction in only two target directions – control audits and cost audits.[2]

However, we believe that control audits will automatically lead to a saving of money, as better control and concretisation of processes will inevitably lead to less room for fraudulent actions or misunderstandings in the awarding, provision of services and invoicing of contracted companies.

And the reverse is also true: cost audits and, if necessary, the discovery of incorrect billing will lead to processes having to be reconsidered and redefined.

Therefore, we basically distinguish between two types of goals that directly influence each other (cf. Fig. 2.2). On the one hand, the goal of improving processes through targeted analysis. On the other hand, the goal of saving costs through targeted control of

---

[2] Risner (2012b).

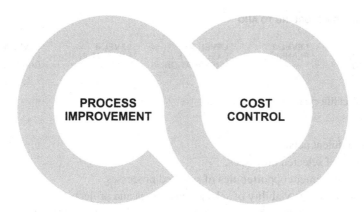

Fig. 2.2 Audit objectives of the construction audit

contracts and invoices. All goals can be divided into further detailed goals or individual goals.

However, this still does not answer the question of how an initiator of a construction audit finds the right objective. Elsewhere, the solution is approached "from behind". The question is posed as to what the project objective is.

Project objectives are the key to risk assessment. In this respect, anything that may prevent the project from achieving its objective constitutes a risk. Thus, all circumstances that could prevent a construction project from being completed in the desired quality, at the planned cost and in the planned time can be risks.

In general, the objectives of the construction audit will always be measured against this principle. The aim of a construction audit will therefore be to identify the greatest risk factors and to make appropriate recommendations for action.[3]

Furthermore, as a matter of principle, objectives must be set for each audit engagement in accordance with the IIA/DIIR. The setting of these objectives must be preceded by an assessment by the internal auditor or the audit manager. The audit objectives must reflect this assessment. Care must be taken to ensure that the defined scope enables the audit objectives to be achieved.[4]

Possible individual goals of the construction audit can be:

- Ensure compliance with laws, policies and internal guidelines,
- Ensure that the deadline targets are met,
- Ensure compliance with building standards and technical regulations,
- Ensure compliance with contracts,
- Ensure that costs are met,
- Ensure the effectiveness of internal control systems,

---

[3] Bünis and Gossens (2016, p. 77).
[4] Bünis and Gossens (2016, pp. 78–79).

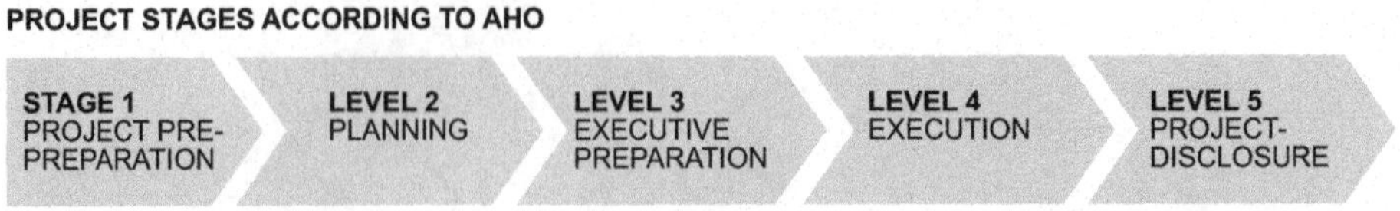

PROJECT STAGES ACCORDING TO AHO

STAGE 1 PROJECT PRE-PREPARATION	LEVEL 2 PLANNING	LEVEL 3 EXECUTIVE PREPARATION	LEVEL 4 EXECUTION	LEVEL 5 PROJECT-DISCLOSURE

Fig. 2.3 Project life cycle

- Detect fraudulent acts,
- Assess risks of selected processes,
- Identify improvement opportunities of selected processes,
- Ensure the economic viability of selected processes and projects,
- Support during the introduction of new processes or instructions for action,
- Repeat testing of processes that have already been tested.

This list shows that audits within the framework of the construction audit can be initiated for completely different reasons and can also be carried out at completely different times in the construction project.

The diagram in Fig. 2.3 schematically shows the project life cycle of a construction process from the initial idea to the finished building. The representation is based on the project stages according to the AHO (see also Sect. 1.3.1).

Each of these project phases involves different participants who work together in different ways. Each project phase conceals different risks that need to be minimized. Thus, the challenge is to find out for each project phase which individual goals are to be achieved and for which goals the greatest risks threaten. Only then does the initiator deal with the concrete object of investigation.

2.3 Test Areas, Test Fields and Test Objects

Section 2.2 noted that the subject matter for the construction audit is primarily dependent on the audit objective to be achieved. Another dependency is on the overall audit planning of the internal audit function. This is determined by the audit map and the prioritisation within the company. With regard to the structure of an audit map, Bünis and Gossens distinguish between five components:[5]

1. Corporate Functions,
2. Business Processes,
3. Business Units,
4. Companies, and
5. Projects.

---

[5] Bünis and Gossens (2016, p. 63).

The tasks of construction auditing are mostly concerned with projects and issues specifically formulated therein or with technical or construction-specific company processes.
Examples of this are:

- Review of supplemental management on new construction project A.
- Review of the warranty management processes in department B
- Review of the documentation of the construction of project C
- Review of the award of sub-threshold planning services in Division D

From my own experience, the determination of the objects of investigation is often done
as follows:

- Determination by management without concretization,
- Determination systematically based on test plan (e.g. test fields in sequence),
- Determination on the basis of concrete cause (suspicion of excessive costs, exceeding
  the construction time, suspicion of fraudulent actions, unclear documentation, suspicion of excessively long processing times, etc.).

However, the intriguing question arises, if no occasion-related construction audit is to take
place, how an initiator of a construction audit can commit to a subject of investigation.

The terms test area, test field and test object do not follow a uniform and generally valid
definition and do not represent protected terms in the sense of a regulation.

Experience has shown that an audit field is understood to be a topic that is coherent in
terms of content or time and can be distinguished from other topics. For example, the
awarding of planning services is a topic that can be delimited from other topics within the
project life cycle of a construction process in terms of both content and time.

Wingsch categorizes construction audits into the following five major audit areas:

1. Technical aspects,
2. Development investment costs,
3. Dates and deadlines,
4. Legal aspects, and
5. Business audits.

The categorization is differentiated here according to the required expertise of the auditor.[6]
The approach is quite understandable, but often leads to problems in the concretisation and
delimitation of the objects of investigation. Let us take the planning contract as an example. According to the initial assessment, this object of investigation requires the examination of legal aspects. However, a closer look reveals other aspects. When it comes to the
question of how it is regulated in the planning contract that and in what form and quality

---

[6]Wingsch (2005, pp. 64–65).

the planner records the construction status or updates the schedule accordingly, other specialist competences become necessary.

Risner summarizes the subjects of investigation, measured by the respective needs of the initiator of a construction audit, as follows:

• Testing the control system,
• Examination of costs,
• Combined control and cost control,
• Review of the building or planning department,
• Change Management Review,
• Review financial and budget reporting,
• Review of construction/architect contracts, and
• Review of supplement management.

The needs of the initiator are inevitably closely linked to its goals. This approach thus supports a goal-oriented investigation in any case. However, no reference is found to the project process and thus not to specific time requirements.

From their own experience, initiators of construction audits often identify these eight objects of investigation:[7]

1. Project Management,
2. Tendering and awarding of architectural and engineering services,
3. Implementation of the planning services,
4. Tendering and awarding of works,
5. Construction,
6. Change management incl. supplement management,
7. Accounting of planning and construction services including all supplements, and
8. Acceptance of design and construction services.

The positive aspect of this approach is that both subject-specific and construction phase-specific objects of investigation are addressed. Our criticism of this approach is that no clear separation of actors and project phases can be identified. For example, the third field of examination, "Execution of planning services", can be both the examination of services up to the completion of the approval planning or only construction supervision services during execution. In particular, the first audit field "Project management" extends from project development to handover of the keys and is thus not very specific.

DIIR Standard No. 4 defines the terms audit area, audit field and audit subject. Accordingly, an audit field represents an independent area. The audit area contains several audit fields that are related to each other, but can basically be audited independently. Audit objects are objects within a audit area and can be processes as well as documents.[8]

---

[7] Schwager and Fischer (2008).
[8] Düsterwald (2010, p. 28).

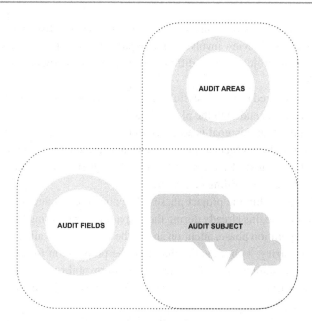

Fig. 2.4 Audit objects as intersection of audit fields and audit areas

This approach shows a certain hierarchy between the three terms. However, one can also represent this approach as a matrix. The test areas can be represented as columns, the test fields as rows of a matrix. Inside the matrix are the audit objects as the intersection of audit area and audit field.[9] In Fig. 2.4, this approach is implemented as a matrix according to DIIR.

This clustering allows a clear delineation in the preparation and planning of an audit. In this way, the audit items can be catalogued and the scope of the audit can be better estimated. Incidentally, this approach can also be found in a similar form for audit planning in other specialist areas.[10]

For the planning and execution of a construction audit, a clustering/categorisation according to three areas has become established in our company. We consider the following as delimitable audit areas:

- Project management (e.g. project management services, project control),
- Planning services (all services that are required according to HOAI),
- Construction (execution of construction work).

[9] Düsterwald (2010, p. 37).

[10] Düsterwald (2010, pp. 37–39).

The background to this is that, as already explained in Sect. 1.3, from a technical point of view, these three areas of review involve different parties who, although they work together within a construction project, have a different view of the construction project from their own point of view.

A project is considered to be successful if the objective is achieved in the planned time, within the planned costs and with the planned quality.

How it is ensured that this goal is achieved depends to a large extent on the organization, coordination and documentation of project development. The review of a project management in the context of the construction audit is therefore significantly concerned with the review of these overriding characteristics.[11]

The risks change within the project phases. The construction auditing department has to react appropriately to this already during the preparation and during the execution of the audits.[12] In the preparation phase, attention should be drawn above all to the probability of technical risks occurring.[13] On the other hand, in the preparation for execution, i.e. after completion of the planning, risks from uncertain approvability, from possible planning errors or faulty construction procedures as well as possible changes in the law will have to be assessed and managed.

Further risks can arise in the choice of the right type of award and with regard to the form of tender. For example, the choice of a tender with individual awards is associated with the risk of having to coordinate a large number of companies. On the other hand, there is the chance of achieving good award decisions and thus saving the budget. A decision to award the contract to only one contractor (general contractor) relieves the project management and provides planning security for the budget, but leads to the risk of having little influence on the execution.[14]

The risks in the execution phase are determined by the change management, by supplements of executing companies, but also by ordered performance changes by the client himself. In addition, there are risks due to coordination errors on the part of the planners or risks from the building ground.[15]

However, the greatest risk in a construction project is changes to the construction design. The later these changes occur in the project, the more risky the realisation of changes is in relation to the achievement of the project objectives. In this respect, a construction auditor will regularly be very concerned with how to ensure that changes are prevented. It is also the task of the construction auditor to check how project management ensures that changes are made with as little risk as possible.[16]

---

[11] Schneider (2017, p. 1).

[12] Schneider (2017, p. 13).

[13] Schneider (2017, p. 16).

[14] Schneider (2017, p. 17).

[15] Schneider (2017, p. 18).

[16] Risner (2012a, p. 114).

In conclusion, this leads to the realization that this periodic risk management in project management must adequately take into account changing risks. For the construction auditor, this means making an assessment of this appropriateness and fundamentally reviewing what means project management uses to respond appropriately to these changing risks.[17]

As test fields we therefore orientate ourselves on the chronological sequence of a construction project or the project stages according to the AHO, as we have also presented it in Sect. 1.3. The relevant test fields are therefore

- Project Preparation,
- Planning,
- Execution Preparation,
- Execution, and
- Project Completion.

In Fig. 2.5, the selected test areas and test fields of the construction audit are related to each other. Project management services will therefore be required throughout the entire project life cycle. This means that audit subjects can arise from all project phases. Construction services, on the other hand, occur at the earliest when execution begins. In this case, audit items will only arise from the last two project phases.

Even if the test fields for different test areas have the same designation, the test objects within these test fields differ for each test area.

Let's take the "execution" check field as an example. If it is a question of checking the preparation of supplements during execution, different processes and risks arise for a company that prepares supplements than for the planner who has to technically check a supplement that has been prepared as part of the project supervision. Other processes and risks arise for an employee from project management, who usually has to evaluate this supplement from a contractual point of view.

Check areas and check fields can also be represented as a matrix for the construction audit. This can be done in such a way that the **test fields** (project stages) are shown as rows

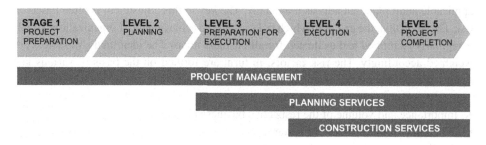

Fig. 2.5 Audit areas in the phases of project implementation (Schneider 2017, p. 5)

---

[17] Schneider (2017, p. 55).

and the **test areas** (service types) as columns in the matrix. The field contents of the matrix form corresponding **test objects** (tasks, processes, documents, procedures).

Table 2.2 contains a proposal for the implementation of the described classification with exemplary contents.

The IIA Standards recommend risk-oriented audit planning by the head of Internal Audit to ensure prioritization. According to the DIIR Standards, suggestions from within and outside of Internal Audit can be taken into account for this purpose in addition to the specific requirements of management.[18]

It is recommended that a risk assessment be performed for the audit areas under consideration. While this recommendation generally applies to the development of an overall audit plan or annual plan for an internal audit function, it is also appropriate for the identification of relevant risk factors and corresponding audit areas for the individual audit.[19]

When planning and preparing for an audit, the DIIR recommends first obtaining an overview of which phase the project is in.

If the project is in an early phase, not many project results are available yet, but plans and planned processes can be checked. The audit results can be used to influence the continuation of the project. The findings from the construction audit flow directly into the construction project.

If the project is far advanced or even completed, the choice of possible audit topics is greater. However, the time between the audit, the handover of the report and the effectiveness of the countermeasures is then relatively short or even non-existent compared to the remaining project duration. The implementation of measures then only has an influence on future projects.[20]

However, since construction projects are unique projects, there is no guarantee that project-specific measures will actually have an impact on other construction projects.

2.4   Project Selection

According to current standards, internal auditing in companies does not follow a generally applicable audit plan, but adjusts it annually. The key to this adjustment is often based on risk. Based on this, the extent to which risks affect the achievement of organizational goals is regularly examined and evaluated for all business areas. The identified risk factors are weighted accordingly. The risk factors, in turn, are formed on the basis of various key figures. The following, for example, can be taken into account:

- Importance and volume of the respective business processes,
- the quality of internal control systems,

---

[18] Berwanger and Kullmann (2012, p. 67).

[19] Berwanger and Kullmann (2012, p. 190).

[20] Düsterwald (2010, p. 31).

Table 2.2 Audit areas and audit fields of the construction audit

	Audit area: Project management	Audit area: Planning services	Audit area: construction services
Audit field: Project preparation	Organization and processes Risk management Timing Internal control systems Process consulting Contract review Building internal competency ...		
Audit field: Planning	Tendering and awarding Planner Economic efficiency calculations/budget planning	Implementation of planning services Deadline control Cost estimation	
Audit field: Preparation for Execution	Change Management Test and control tasks Tendering and award of contract Construction ...	Tendering and award of contract Construction Organization preliminary measures Budget compliance ...	
Audit field: Execution	Control of costs, deadlines and quality Test and control tasks Information and communication Contract Management ...	Construction supervision Control of costs, deadlines and quality Test and control tasks Change Management Supplementary audit Correspondence management ...	Construction documentation Construction execution Deadline monitoring Claims management Correspondence management Dimensions Accounting ...
Audit field: Project completion	Documentation Test and control tasks User handover ...	Revision planning Final invoices Acceptance Warranty management ...	Documentation requirements Final invoices Acceptance Construction sites - clearing ...

- Complexity of the individual processes,
- Dynamics of the audit objects,
- Composition of the personnel employed, and
- Periods of the last examinations etc.

A risk score is then derived from a complex assessment using a large number of test algorithms. The risk score is used to determine which processes or business areas are included in the audit plan.[21]

Projects and especially construction projects are – in contrast to organisational units or repetitive processes in organisations or companies – particularly risky because of their uniqueness.[22] Criteria that can be considered for the selection of construction projects to be audited are, for example, similar to those shown in Fig. 2.1 on (please generally no page references):

1. Project Size,
2. Scope of Change,
3. Project complexity,
4. Contract model,
5. Budget Compliance,
6. Deadline Compliance,
7. Planning quality,
8. Completion of planning,
9. Fault Scope, and
10. Novelty.

Similar guidance for project selection can also be found in the American literature. Risner approaches the selection of his review project for a construction review using a risk matrix. He suggests scoring the eligible projects on certain attributes and then comparing them. The project with the highest score contains the greatest risks and thus becomes the target of the audit.

Following Risner, we propose to evaluate the following 10 attributes for construction projects. A maximum of five points are awarded for each attribute.[23] The evaluation can – adapted to German conditions in the construction industry – look as follows.

1. Project Size

The project size can be determined by the total investment volume, the sum of the costs from KG 300 and 400 according to DIN 276 or, for example, by the volume of the division

[21] Bünis and Gossens (2016, pp. 52–58).
[22] Düsterwald (2010, p. 24).
[23] Risner (2012a, pp. 37–45).

that is being examined. Alternatively, a differentiation in project durations could be made or a performance density could be used as a differentiation criterion (e.g. measured in EUR per month). The idea behind this criterion is that problems in a project with a high investment volume inevitably also entail high financial risks.

In Table 2.3, the project size has been evaluated in terms of the total investment volume.

2. Scope of Change

Deviations from the construction design regularly mean that plans have to be changed, materials have to be re-sampled or re-ordered, labour has to be re-coordinated, processes may have to be re-arranged, etc. All of these effects lead to the risk that costs and the construction schedule cannot be met or that qualities change. All these effects lead to the risk that costs and construction time cannot be met or that qualities change. The expression of deviations can be the increase of the order volume.

A better indicator, however, is the number of changes. Even very small, relatively inexpensive changes can have an enormous impact on the project as a whole. For construction projects, the number of supplements is therefore a good indicator. The changes are defined by cost changes in the construction volume in relation to the total construction volume as a percentage. Alternatively, the number of (subsequent) contractor amendments could also be taken into account, as shown in Table 2.4 as an example.

Table 2.3 Assessment of project size

Not more than EUR 5 million	1 points
More than EUR 5 million and not more than EUR 10 million	2 points
More than EUR 10 million and not more than EUR 50 million	3 points
More than EUR 50 million and not more than EUR 100 million	4 points
More than EUR 100 million	5 points

Table 2.4 Assessment of the scope of change

Not more than 25 supplements	1 points
More than 25 and not more than 50 supplements	2 points
More than 50 and not more than 100 supplements	3 points
More than 100 and not more than 500 supplements	4 points
More than 500 supplements	5 points

3. Project Complexity

The more complex a project, the more diverse the risks associated with the project. For example, the conversion of an existing clinic and the continuation of its operation will be considerably more complex than the construction of a new warehouse for the municipal building yard in an open field.

Similarly, a collaborative construction project with multiple stakeholders (e.g., project with participation by a municipal corporation, the respective state, different funding agencies) will be more complex than a construction project with only one stakeholder (e.g., project by only one municipal corporation).

In the following, the complexity is assessed via the degree of requirements with which the project must be planned, implemented and coordinated (cf. Table 2.5). In accordance with the HOAI, either individual aspects can be evaluated, such as requirements for integration into the environment, number of functional areas, design requirements, construction requirements, technical equipment or finishing.[24] Or the evaluation is carried out on the basis of a list of objects.[25]

4. Contract Model

The decision for a certain type of construction contract is essential for the assessment of the risks of the construction project. Although the unit price contract with the individual award of all construction services may place high demands on the project management, this type of contract is nevertheless to be assessed as low-risk, since the tenders are made item by item and usually in great detail. In addition, only those services are invoiced which have been demonstrably performed.

Construction contracts that are based on a lump sum are more risky. Even if the quantity risk is passed on to the contractor, the client loses control options. Furthermore, there

Table 2.5 Assessment of project complexity

Very low complexity (e.g. simple temporary buildings, simple residential buildings)	1 points
Low complexity (e.g. simple closed, single-storey halls)	2 points
Average complexity (e.g. shopping malls, hotels, large residential buildings)	3 points
High complexity (e.g. railway stations, groups of houses in dense construction)	4 points
Very high complexity (e.g. airports, university hospitals, power plants)	5 points

---

[24] Cf. §35 HOAI – Fees for basic services for buildings and interiors.

[25] E.g. based on Annex to HOAI 10.2 – Object list buildings.

Table 2.6 Evaluation of the contract model

Unit price contract with individual award	1 points
Unit price contract with general contractor	2 points
Detailed lump sum price contract	3 points
Global lump sum price contract	4 points
Other contract type (e.g. cost reimbursement contract, hourly wage contract or similar)	5 points

Table 2.7 Assessment of budget compliance

Budget respected or exceeded by max. 10%	1 points
Budget exceeded by max. 25	2 points
Budget exceeded by max. 50	3 points
Budget exceeded by max. 100	4 points
Budget exceeded by more than 100	5 points

is the risk that services are not adequately described in the underlying subject of the tender or have been forgotten. Changes or additions become expensive afterwards.

The greatest risk is associated with the agreement of a cost-reimbursement amount, since neither costs nor time can be calculated for this type of contract.

In Table 2.6 we have evaluated for construction projects the distinction whether individual awards or general contractor contracts, lump sum contracts, unit price contracts, etc. have been agreed.

5. Budget Compliance

The adherence to the budget or the deviation from the planned budget – without changes to the construction design – represents an important indicator for the review of a project. Significant overruns of the budget indicate that there have been significant problems during project execution, which need to be analysed. These problems indicate which risks have actually occurred.

In Table 2.7, budget compliance is assessed as a relative change to the agreed order volume. The lowest risk exists in the case of deviation within 10% (oriented to §2.3 VOB/B). If the order volume more than doubles, the project receives the highest score.

6. Adherence to Deadlines

Adherence to agreed deadlines is another key indicator for construction projects. If an agreed schedule is exceeded, it may be because a contractor simply did not have sufficient capacity. In this case, the contractor's capacity may not have been adequately checked during the tendering and award process. Another reason, however, may be that the project management did not prepare a proper schedule and update it accordingly. In any case, it

Table 2.8 Assessment of compliance with deadlines

Schedule adhered to or exceeded by max. 10%	1 points
Schedule exceeded by max. 25	2 points
Schedule exceeded by max. 50	3 points
Schedule exceeded by max. 100	4 points
Schedule exceeded by more than 100	5 points

makes sense to review projects that show significant schedule overruns in order to determine which risks have materialized in these projects.

In addition, very long-term construction projects also involve high risks, as the personnel involved frequently change, legislation can change and companies can go bankrupt.

Table 2.8 evaluates the adherence to deadlines measured by a relative overrun as a percentage of the originally agreed schedule. If the construction time more than doubles, the project receives the highest score.

7. Planning Quality

Poor quality planning has a variety of effects. It will lead to cost calculations not being correct. It will further lead to the tenders for the construction work also being of poor quality and the execution plans possibly being non-functional. If this occurs, rework will be required which will result in addenda. If this occurs, rework will be required, which will result in supplements. In addition, improvements to the planning regularly lead to considerable time delays.

Evaluating the quality of planning is a great challenge, which is difficult to measure. In addition, the measurement result should also be comparable with other projects. This is already very difficult because every construction project is unique and brings different challenges with it.

In Table 2.9, the quality of the planning is therefore assessed on the basis of the number of objection notifications, as required in accordance with §4.3 VOB/B. A high number of objection notifications shows that the planning of the construction project was obviously not technically correct and unambiguous enough to allow the contractor to build without doubt. In the following table, the number of concerns has been graded in a range of 10–100 concerns. This classification can be selected individually according to the size of the project.

For example, if projects are compared that are all very complex and have a high contract volume, it is likely that all projects will have a high number of concerns. The opposite is true when comparing very small and simple projects.

Table 2.9 Assessment of planning quality

No more than 10 notifications of concern	1 points
More than 10 and not more than 25 notifications of concern	2 points
More than 25 and not more than 50 notifications of concern	3 points
More than 50 and not more than 100 notifications of concern	4 points
More than 100 concerns raised	5 points

Table 2.10 Assessment of the completion of planning

Implementation planning 100% complete	1 points
Final design at least 80% and less than 100% complete	2 points
Final design at least 60% and less than 80% completed	3 points
Final design at least 40% and less than 60% complete	4 points
Implementation planning less than 40% complete	5 points

8. Completion of the Planning

It is not unusual nowadays for the execution of construction projects in Germany to begin before the complete execution planning is available.

For example, it may be that the shell of an office building has already been started, while the implementation planning of the building services is still being processed and, for example, air flow measurements have not yet been completed. This can lead to the ventilation ducts having to be dimensioned larger than previously assumed. This in turn may mean that ceiling heights can no longer be maintained or that ventilation ducts collide with other building services lines. If this occurs, the construction costs will increase enormously and the construction time will inevitably be exceeded.

In Table 2.10, the degree of completion of the implementation planning (according to LP 5 HOAI) before the start of construction is evaluated in percent.

9. Scope of Malfunction

If a construction project is disrupted for a variety of reasons, there will inevitably be changes to the construction sequence, delays in the construction process, additional

Table 2.11 Evaluation of the scope of disturbance

No more than 25 obstruction notices	1 points
More than 25 and not more than 50 obstruction notices	2 points
More than 50 and not more than 100 obstruction notices	3 points
More than 100 and not more than 500 obstruction notices	4 points
More than 500 obstruction notices	5 points

Table 2.12 Assessment of novelty

Standard project, which has already been realized numerous times in this way	1 points
Project, which has already been realized several times (more than 10 pcs.) in this way	2 points
Project that has already been realized in this way in isolated cases (no more than 10 pcs.)	3 points
Project already realized in a comparable way	4 points
One-of-a-kind, the first of its kind	5 points

acceleration measures or even extensions to the construction period, etc. All these effects affect the adherence to the costs and the agreed construction time.

Table 2.11 evaluates the number of obstruction notices as required by §6.1 VOB/B. A high number of obstruction notices shows that there are obvious reasons in the construction project which prevent the contractor from building without doubt and according to plan. In the following table, the number of obstructions has been graded in a range of 25–500 obstruction notices. This classification can be selected individually according to the size of the project.

10. Novelty

Even though construction projects are unique, there are numerous construction projects that have already been implemented several times in the same or similar form by a project team or a company. The best known representatives are building projects in system construction or modular construction. These can be, for example, gymnasiums or day care centers.

In such projects, the same plans are implemented several times at different locations. In some cases, even the same companies are involved in the execution. Experience from previous projects, an experienced project team that has been carrying out such projects for years is less risky than a newly assembled team.

In contrast, the construction of a lookout tower with an unusual design will be a unique project that involves many unknown risks. Table 2.12 evaluates whether the project has already been realized many times or is the first of its kind.

In Chap. 3, examples are given to show how this matrix is used.

2.5   Basic Determination

The preparation of the audit – i.e. the basic determination – is a central success factor for every audit assignment. Starting from the objectives and the definition of the test object as well as the test items, all necessary steps have to be organized. This includes in particular the selection of the right contact persons and the collection of the right information.

In addition, the actual audit procedures must be prepared, such as the formulation of relevant questions and, if necessary, the performance of initial analyses. Here, for example, the information from the final account – if already available – can be used to carry out initial key figure comparisons. Or deviations identified during repeat audits can be compiled.[26]

A construction auditor will have an overview of the internal control tools and related documents before starting the audit. Nowadays, these will mainly be electronic elements. This preparatory phase includes

- Formal evaluation of the documentation, e.g. by rough analysis of the project file, browsing the directories, quantitative evaluation of the contents,
- Review of relevant documents with a view to including issues in the audit programme; and
- Recording and representation of processes in order to gain an understanding of them and to formulate questions.[27]

Often, once the audit objective, project and audit subject have been identified, the necessary participants and the documents and information required for the audit almost arise by themselves. But here again, an algorithm can assist with the set-up.

Selection of Stakeholders
A suggestion for the necessary participants can be determined for a systematic approach from the matrix presented in Sect. 2.3. For example, the management, the project manager and the contracting authority will be important discussion partners for the review of the project management audit area during project preparation. This is because, among other things, clarification is required in this project phase,

- such as project objective, project budget and time schedule are defined,
- whether a project controller will be used,
- which types of award are eligible,
- which stakeholders need to be involved,
- what funding, if any, is available,
- which specific project risks exist,

---

[26] Eulerich (2018, p. 244).

[27] Bünis and Gossens (2016, pp. 81–82).

- how the project team should be organized, etc.

On the other hand, the site manager, coordinator for safety and health matters[28] or contract manager will be particularly relevant for the audit of the area of construction work during execution. This is because, among other things, clarification is required in this project phase,

- how the construction progress is checked and documented,
- whether and how the project schedule is properly updated,
- how safety is checked on the construction site,
- how faults and defects are documented,
- whether and when necessary interim audits take place,
- how to carry out economical construction site logistics
- whether and how the construction companies' supplements are checked, etc.

Table 2.13 uses the matrix presented in Sect. 2.3 to show the possible relevant parties involved depending on the test area and test field. The presentation does not claim to be exhaustive. For very large projects, there may be many more relevant stakeholders. In contrast, for very small projects, only one contact person may be relevant for all issues.

Selection of Information
In addition to the interviews with the specific project participants according to Table 2.13, the examination requires an insight into certain documents. These can be documents (such as contracts, minutes or forms) as well as information on types of procedures (process diagrams, procedural instructions, etc.).

A suggestion for the required information and documents can also be determined for a systematic approach from the matrix presented in Sect. 2.3. For example, the audit of the following documents will be relevant for the review of the audit area project management during project preparation:

- First project outline, derivation of project objective, project application,
- Procedural instruction and process flow of the project development,
- Budget plan and schedule with justification,
- Tender documents and process for tender project controller,
- Retrieval Requirements of Stakeholders,
- Risk matrix for project development,
- Organizational chart project team with role description etc.

On the other hand, the audit of the following documents will be relevant for the audit of the construction work during execution:

---

[28] "SiGeKo" in German.

Table 2.13 Stakeholders to be involved in the audit

	Test area: Project management	Test area: Planning services	Test area: construction services
Audit field: Project preparation	Management Project management Awarding body …		
Audit field: Planning	Management Project control Project management …	Project management Planner …	
Audit field: Preparation for execution	Project control Project management Awarding body Contract Management …	Planner Project Management Contract Management …	
Audit field: Execution	Project control, Project management, Contract management …	Planner Project Management Contract Management …	Site management Project management SiGeKo, Contract management …
Audit field: Project completion	Project management Contract management Invoicing office …	Planner Project Management Warranty Office Invoicing office …	Construction management, Project management, Warranty office, Accounting office …

- Process instruction and process flow of the construction site documentation,
- Construction site documentation and daily construction reports
- Updated project schedule,
- Minutes of the coordinator for safety and health matters,
- Fault and defect documentation including corresponding process instructions,
- Protocols of necessary interim audits,
- Review process for supplements with examples etc.

Table 2.14 uses the matrix presented in Sect. 2.3 to show the possible relevant parties involved depending on the test area and test field. The presentation does not claim to be exhaustive. In the case of very large projects, significantly more processes and documents

Table 2.14 Documents required for the test

	Test area: Project management	Test area: Planning services	Test area: construction services
Audit field: Project preparation	Requirements planning Area program Utilization program Risk assessment Organization chart Project manual ...		
Audit field: Planning	Assignment of planning Budget plan Rough schedule Minutes ...	Planning contract Tender documents Planners Planning documents Scheduling Cost calculation ...	
Audit field: Preparation for execution	Contract award process Construction execution Documentation Review of utilization program Rough schedule Budget update ...	Tender documents construction companies Plan documents Documentation awarding processes Timing ...	
Audit field: Execution	Contract review Documents and processes Contract management Detailed schedule Budget update ...	Documents and processes Subsequent management Scheduling Cost control Logging Correspondence management ...	Construction site documentation SiGeKo protocols Measurements Correspondence management Settlements Supplements ...
Audit field: Project completion	Final invoice audits	Planning contract Revision plans Acceptance records	Final account preparation Acceptance protocols
	Handover protocols to users ...	Final invoice audits Defect management Documentation Warranty Cost determination ...	Handover warranty Documents according to documentation requirements ...

may have to be audited, which is why a further restriction based on the audit subject to be audited becomes necessary.

After the selection of the parties involved and the relevant documents, as well as their retrieval and initial audit, an audit programme is to be developed and set up.

At this point, it is important for the development to assess whether a full audit will be required or whether a partial audit will already provide sufficient information to achieve the objective of the audit.[29] This assessment will have a significant impact on the required schedule.

According to IIA/DIIR, the audit programme must contain all essential procedural steps and must be approved before the audit begins. Approval must not be given by the construction auditor himself or a person from the audit team, but by the head of Internal Audit.

If the internal auditors plan to have the construction audit performed by external auditors – which is very often the case due to the special expertise required – a written agreement must be reached on the expectations and objectives as well as the responsibilities and access rights.[30] This preparation therefore includes, on the one hand, drawing up a timetable (see Sect. 2.6) and, on the other hand, defining meaningful questions.

Questions that are already answered by the documents received can be neglected. Questions that do not apply to the test field or object can also be neglected.

For example, various questions on the award process are superfluous if, due to the threshold values being exceeded, all award steps are regulated via a certified public online portal with defined processes anyway.

A suitable possibility is the use of the so-called MOPTIC method (Medium/Organisation/Performance/Time/Control – in short: MOPTIC).[31]

With this method the quality of processes, organizations and documents can be checked in the same way.

Regardless of the test object is questioned:

- Medium (the "how" question),
- Organization (the "who" question),
- Performance (the "how well" questions),
- Time (the "when" question),
- Control (the "if" question).

Simplified, however, it is also possible to work with a firmly defined questionnaire. The answers are recorded in a database on a project-by-project basis. During repeat tests, the results can be compared with each other and the success of improvement measures can be checked.

---

[29] Berwanger and Kullmann (2012, p. 203).
[30] Bünis and Gossens (2016, pp. 83–84).
[31] Düsterwald (2010, p. 98).

In addition, a database offers the possibility to record all audits. Thus, it is possible to benefit from the experience of previous audits, e.g. by adopting appropriate measures from other audits.

After it has been shown how the relevant parties involved and the corresponding documents and information can be systematically located using the matrix of test areas and test fields, it quickly becomes clear that it is also appropriate to compile the essential questions using the matrix. It is recommended that the compilation and management of these questions be carried out with the aid of IT checklists.

Excel tables or the construction of a simple database are best suited for this. However, complex databases are better suited for regular and repetitive audits, which may also allow an evaluation of the mass data. This can even go so far that such databases already provide essential parts for reporting at the end of an audit.

As a result, an audit can be carried out in a time-saving manner and a lower error rate can be assumed. In addition, the evaluation can also take place significantly faster and with consistent quality.

When audits can be performed more quickly and with less effort, there is more time for a construction auditor to be creative in compiling and discussing recommendations and coming up with possible actions to improve construction processes.

As soon as it is possible to carry out construction audits in a shorter period of time, recurring and of better quality, with less effort and therefore lower costs, there is also the possibility of carrying out audits more frequently. More frequent audits, in turn, lead to greater awareness among all stakeholders and thus can lead to greater discipline in the execution of construction projects. In addition, regular and frequent audits will lead to an improvement in the quality of the audit.[32]

In addition, recurring systematically prepared audits based on the same participants, the same documents and the same questions will lead to an increase in the maturity of the construction processes.

2.6   Scheduling

In principle, a construction review can take place at any time. The decision as to when a construction review should take place depends on the nature and objective of the review, the nature of the construction contract and the resources available for a review.[33] The type and timing of the audit is essential to the scheduling of the audit.

A project-accompanying audit (**audit carried out ex-ante**) requires a continuous deployment of the construction auditor over the entire project period. Within this assignment, interim reports will help to improve certain construction processes and influence

---

[32] Berwanger and Kullmann (2012, pp. 207–209).

[33] Risner (2012a, p. 73).

further project processing. The objectives can thus be both process improvements and cost savings.

However, care must be taken in such audits that the constant presence of the construction auditors over the entire project duration requires a high audit capacity. In addition, care must be taken to preserve the independence of the construction auditors. Often their mandate is misunderstood as a consultancy service and, because of the permanent proximity to the project, the construction auditor runs the risk of becoming part of the project team himself.[34] The assignment times and the scheduling of the building auditor will be strongly measured (or oriented?) by the project schedule.

A one-time and retrospective review of a single or multiple audit fields (**ex-post audit**), on the other hand, will take place contiguously over a specific period of time. This period is independent of specific events in the project. If necessary, the project may already have been completed when the review process begins. The main purpose of the findings is to improve the processes in future projects. As a rule, the check can no longer contribute to cost savings in the project concerned, since completed actions are checked. The deployment times and the scheduling of the construction auditor are independent of the project schedule.

The scheduling of an event-related review of a selected process (**ad hoc audit**) will also take place contiguously within a certain period of time. Such audits – which regularly occur in cases of fraudulent actions – are difficult to predict in terms of content or time with regard to implementation and duration.[35]

However, the aim will be to carry out this audit close to the time of the project in order to have an influence on upcoming project steps if necessary or to ensure access to the participants who have not behaved properly in this project. The objectives can thus be both process improvements and cost savings.

Even if the three types of audits presented appear to be very different, the same essential steps basically apply to the execution of the construction audit. The audit is understood as a process in which each step of the process represents a necessary, self-contained section, which in turn can be structured individually.

Adherence to the uniform basic structure of the audit process ensures that all audits are based on identical framework conditions and criteria. This makes it possible to present a uniform image to the audited entities and to improve quality assurance.

The main process steps are shown in Fig. 2.6. Accordingly, a construction audit can be divided into five essential process steps like any other revision process.

1. **Planning**

The planning of the audit process takes place as a specification of the terms of reference of the audit assignment within the framework of a personal discussion between the initiator

---

[34] Düsterwald (2010, p. 102).

[35] Berwanger and Kullmann (2012, p. 195).

Fig. 2.6 Auditing process

of the construction audit and the auditor. In this phase, the objectives are clarified and the test object and test items are defined. Ideally, these specifications are made in connection with a risk assessment. In addition, an exchange of initial project-specific documents and information can take place in order to outline the required deployment of staff, obtain any necessary approvals and determine a rough schedule.

2. **Preparation**

The preparation of the audit process includes familiarization with the project documents received in advance and the preparation of an audit plan. The audit plan can already distinguish which information the construction auditor can obtain from the audit of the documents and for which information interviews with project participants are unavoidable. In addition, the audit plan can already contain a list of the relevant questions for the specified audit objects. Likewise, the audit plan will further differentiate the rough sequence in terms of time.

Subsequently, the construction auditor will prepare a kick-off meeting. The aim of the kick-off meeting is to get all the necessary parties involved (see Sect. 2.5) around the table and, within this framework, to present the audit objectives and the audit plan as well as to further specify the rough schedule together. In addition, the audit objective can be specified if necessary.

In many companies, the performance of a construction audit is characterized by the fact that external construction auditors are called in for this service. In internal auditing, this form of support is also referred to as co-sourcing.

As a result, the participants in the audit usually do not know each other. For this reason alone, a kick-off meeting is regularly recommended at the beginning of the audit. The meeting of all those involved in the audit ensures that uncertainties about the forthcoming audit can be removed. It is particularly important for an external construction auditor to be able to build up trust in the company and department to be audited right at the start of the audit.

The construction auditor is given the opportunity to present his approach to all parties involved and to agree on the time schedule. It is also a good opportunity to arouse interest in the audit and to make the tasks and roles of the individual participants transparent. As a rule, it is not until the kick-off meeting that it becomes clear how long the entire audit will take.

In addition to these overriding objectives, such an interview naturally also serves to request or obtain further documents relevant to the examination from the respective parties.[36]

It is advisable to keep a record of the initial meeting in order to record any changes that may have been made to the protocol or any concerns that may have been raised by the individual parties.[37]

3. Implementation

During the performance of an audit, a construction auditor will involuntarily and compulsorily follow a certain pattern. In the literature, this is also referred to as an audit action (Fig. 2.7).

The construction auditor will first familiarise himself with the circumstances and the object of the audit. This includes independently reading the object description of the construction project, reading the more important contracts, becoming familiar with the project drive or a cloud-based project drive and the documentation on the construction project stored therein, gathering the applicable procedural instructions and processes for the test object and, if necessary, a site audit.

The next step is to check whether the processes and procedures defined in the company or by guidelines are actually to be found in the prescribed manner in the construction project. If necessary, this verification must be supplemented by information that the construction auditor obtains through targeted interviews with the parties involved. These interviews

---

[36] Berwanger and Kullmann (2012, pp. 202–203).

[37] Bünis and Gossens (2016, p. 87).

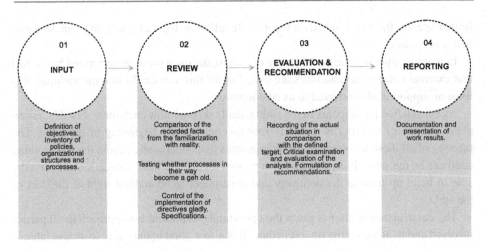

Fig. 2.7 Audit procedures (Berwanger and Kullmann 2012, p. 206)

can also represent on-site audits with independent audit of documentation or construction site situations.

Subsequently, the actually established procedure but also the prescribed procedure must be critically examined with regard to the achievement of the objectives.

For any findings, a construction auditor is advised to provide evidence on which his finding is based. This can be concrete documents, which either confirm rule-compliant behaviour or prove the opposite. Notes on statements from an interview are also suitable as evidence in the event that no suitable documents are available to prove the findings of the construction auditor.

Documentation is important to guard against the pretense that an audit and finding of noncompliant behavior are made based on subjective perceptions. In addition, tangible evidence provides a good opportunity to enter into a possible later in-depth discussion of the audit findings.

An assessment with a corresponding recommendation should be made to answer this question from the audit plan. The recommendation should indicate whether action is required to improve a process or reduce costs.

A multi-level approach is often used. In this context, multi-level means that the recommendation is measured against the risk that arises if the recommendation is not followed.

For simplicity, a 3-tier model is recommended:

Stage 1	No appreciable risk – no measures required
Stage 2	Low risk – recommendation of appropriate measures
Level 3	High risk – recommend urgent action

All findings which have been assessed with level 3 and for which urgently required measures are recommended, it is recommended that the measures are also concretely proposed by the construction auditor.

4. Report and Closure

The conclusion of an audit may take the form of a final meeting or a final presentation. The construction auditor will present the results of his work to the initiator of the construction audit or to the management in a suitable form. Such a final meeting serves to show the results of the audit, but also to clear up disputed points or ambiguities and misunderstandings before the final report is prepared.

After a final meeting, a summary report is usually prepared. This report once again describes the audit objective and the selected procedure. In doing so, it is advisable to revisit all the joint stipulations between the initiator of the construction audit and the construction auditor regarding the content-related and qualitative stipulations.

In relation to these specifications, the audit result is communicated and a catalogue of measures is handed over, which should describe the content, scope and timeline for the implementation of the measures. It is recommended to attach all protocols and evidence for the audit in the annex to the report. For the concrete form of the report, please refer to the explanations in Sect. 4.2.

5. Audit

The follow-up (also called "follow-up procedure") involves checking whether the recommended measures from the audit report have been properly implemented. This can go so far that the effectiveness of the recommended measures is checked again in the form of a follow-up audit at an appropriate time interval.

This activity should be carried out by the construction auditor, as the independence of the assessment is guaranteed by him. However, the follow-up can also be carried out by the initiator of the construction audit himself, as he only has to follow the procedure in the audit report and no new audit planning of his own is required.

The review is often not a fixed component of a construction audit. This may be due to the fact that construction projects are unique and there is no need to carry out a follow-up audit for the same project after its completion.

For example, the objective of an ex-post audit is either process improvement or cost control. A review of process improvements can only be carried out on the basis of later suitable projects. A review of cost control is not relevant, as the objective is achieved with the detection of deviations.

For ad hoc audits, on the other hand, there is no need for follow-up audits, as these usually pursue the goal of uncovering fraudulent acts. With the detection of fraudulent acts, the objective is achieved.

Only in the case of audits carried out ex ante is there a need and also the possibility to carry out a follow-up in the same project, that the objective of these audits is to make improvements during the execution of the project.

2.7   Summary

The essential steps for planning a construction audit, as shown in Sects. 2.1, 2.2, 2.3, 2.4, 2.5 and 2.6, can be summarised in the following algorithm:

• Determine *audit objective* (see Sect. 2.2)

A distinction is made between (1) cause-related audits, e.g. detection of fraudulent acts, and (2) prophylactic audits, e.g. process improvement or cost control. If (1), then project selection and selection of the audit field are clear, if (2), then continue with the next steps.

• Identify the *test area, test field and test object* (see Sect. 2.3).

Narrowing down the scope of the audit using a matrix with audit areas in the columns, audit fields in the rows and audit objects as the intersection.

• Make *project selection* (see Sect. 2.4).

Evaluation of a selection of projects using a risk matrix based on 10 typical risk criteria for construction projects. The project with the highest score becomes the audit project. This step is omitted for event-related audits.

• *Determine* the *basis for* preparation (see Sect. 2.5)

Identify the required participants and the required information using a matrix with check areas in the columns, check fields in the rows, and the participants or the information as the intersection.

• Carry out *scheduling* (see Sect. 2.6)

Following generally applicable recommendations for the implementation of an auditing process.

This proposal for systematic planning of a construction audit is illustrated in Fig. 2.8.

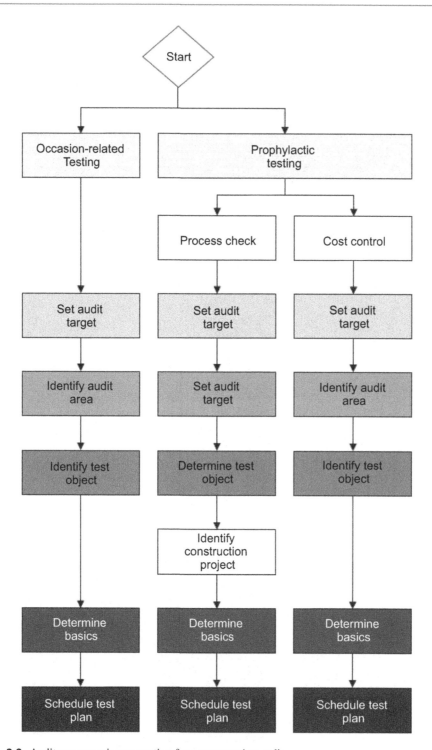

Fig. 2.8 Audit processes in preparation for a construction audit

Fig. 5.8

Implementation of the Construction Audit

3

3.1 Introduction

There is no generally applicable binding specification for the performance of the audit procedures of a construction audit. Chapter 2 showed how an audit programme can be drawn up. It is advisable for a construction auditor, an audit firm or a corresponding department to define its own audit programme.

This ensures that different auditors always carry out their audit in the same quality for different objects. In addition, audit results become comparable. In addition, it is possible to make the success of improvement measures measurable and thus visible during repeat audits.

Section 2.5 outlined the possibilities for this. This chapter uses examples to show how audits can be performed for the three audit cases presented (event-driven audit, prophylactic process audit, prophylactic cost control). Three different case studies are presented. The examples originate from the practice of the authors, but have been anonymized for this book, partially modified and greatly reduced in scope.

3.2 Example of Occasional Auditing

3.2.1 Initial Situation

A construction company, as a total contractor, is taking on the planning and construction services for a shopping centre in a city centre location. The scope of the construction work includes not only the new building but also the partial demolition of an existing building. A global lump sum price contract has been agreed with the client. The construction work will be awarded as partial lots in unit price contracts to subcontractors with whom the VOB is agreed.

In the course of execution, the general contractor finds that the invoiced quantities and masses in some partial lots extensively and in large numbers exceed the front rates of the bill of quantities. In other partial lots, however, the invoiced services correspond sufficiently precisely to the tendered scope and do not deviate by more than 10% from the front rates of the bill of quantities.

This may indicate faulty planning that was corrected in the course of execution. Special attention must then be paid to the services actually performed and their correct billing. If the services have been correctly determined and invoiced, the difference to the front rates of the bill of quantities can be clearly determined. These quantity increases would not be chargeable to a contractor, but would have to be remunerated.

However, it is not always possible to subsequently check the services actually performed. This applies in particular to services that can no longer be determined retrospectively with a simple measurement because they are no longer visible, accessible or present. This includes for example

- Search shafts,
- Earthworks,
- Installations in the ground and in concrete,
- Reinforcement work,
- Winter Construction Measures,
- Cleaning and clearing work, and
- Personnel deployment to secure the construction site.

For these services it is therefore necessary that the type and scope are already documented during or immediately after completion of the work and confirmed by the client.

In the case of reinforcement work, it is therefore common practice to remove the placed reinforcement before concreting. In the case of earthworks, it is necessary to accurately record excavated earth masses and document their whereabouts – but this does not apply to earth masses that remain on the construction site.

In the case of the other work mentioned, too, there is often scope for design or no concrete and binding documentation requirements. Cleaning and clearing work, for example, is often commissioned and remunerated on an hourly basis because of the low proportion of material costs involved. For billing purposes, "wage timesheets" are prepared in which the type and scope of the work performed and the deployment of personnel are documented on a daily basis.

Daily audit and approval by the client's construction supervisor is an important prerequisite for invoicing. If this audit is not carried out daily and conscientiously, the service actually provided cannot be measured.

According to a study by management consultants Deloitte, around 89% of all Germans will own a smartphone in 2020. According to the study, the proportion is even higher among people of working age. These devices are equipped with powerful cameras, Internet access and GPS. Photos taken with these devices are created with exact information on the time and location of the shot and – when handed over to the client in a timely manner – can be a valuable addition to the documentation of the services provided.

3.2.2   Preparation

Thus, an event-related audit is to be carried out. In Fig. 3.1, the audit is classified in the process overview for the preparation of the construction audit and the procedure is followed according to the left column.

Audit Objective

The total contractor commissions a construction audit with the aim of clarifying the quantity overruns in a responsible manner: are there actually planning errors or have subcontractors manipulated the invoicing? Can the documentation clearly show the reasons for the deviations so that the client can be reimbursed for them? Has the construction supervision sufficiently fulfilled its audit, coordination and information obligations? Does this give rise to liability or damage claims? Are criminal, civil or labour law measures to be initiated?

Audit Area

It follows from the objective that the investigation should relate to the audit area of construction services.

Audit Field

It follows from the initial situation that the investigation should relate to the audit field of execution.

Audit Item

The intersection of audit area and audit field results in a selection of relevant audit items, as shown in Table 3.1.

Based on the problems described in the initial situation and the objective – responsible clarification of the quantity overruns – the audit items from Table 2.2 can be specified as follows:

- Auditing contracts of designers and performers,
- Auditing the processing and settlement of supplements,
- Auditing the documentation of the supplementary audit,
- Auditing costs and impact of supplements on budget,
- Auditing measures for early detection of deviations.

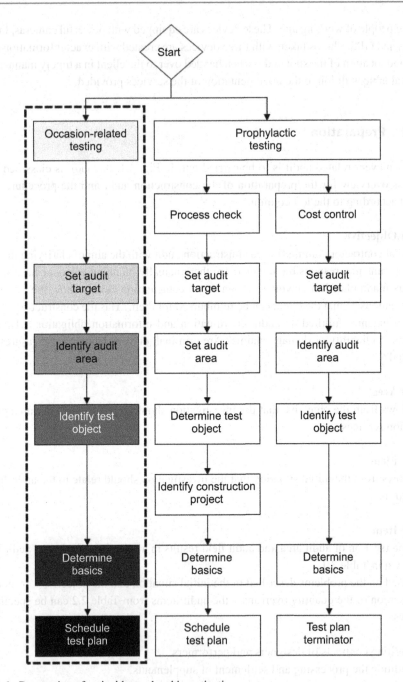

Fig. 3.1 Preparation of an incident-related investigation

Table 3.1 Delimitation of audit area and audit field Example 1

	Test area: Project management	Test area: Planning services	Test area: construction services
Audit field: Project preparation			
Audit field: Planning			
Audit field: Preparation for execution			
Audit field: Execution			
Audit field: Project completion			

Once the audit items have been identified, the basis required for the audit shall be determined.

Basic Determination

The audit is mainly based on two types of sources. Firstly, the audit is conducted as a literature search. In this context, literature is understood to be all documents provided by the company. This takes place, for example, in the form of access to an electronic project folder or a web-based document management system.

On the other hand, the findings from the literature study are verified and supplemented by expert interviews. In this context, experts are understood as those persons who are involved in the execution of the project and can contribute significantly to answering the question through their specialist knowledge.

With the help of the selection matrix from Table 3.2, it is possible to derive from the intersection of the audit area and audit field which of the parties involved can be considered for the expert interviews. Based on the problem described in the initial situation and the objective – responsible clarification of quantity overruns – the participants from Table 2.13 can be specified as follows:

- Project Management,
- Site Management,
- Project Management,
- Construction Supervision,
- Legal Department,
- Purchasing,
- Coordinator for safety and health matters, and
- Contract Management.

Table 3.2 Delimitation of audit area and audit field Example 2

	Test area: Project management	Test area: Planning services	Test area: construction services
Audit field: Project preparation			
Audit field: Planning			
Audit field: Preparation for execution			
Audit field: Execution			
Audit field: Project completion			

It should be possible to substantiate the responses of the parties involved by means of supporting documents. However, they can also be expressed as a statement and must then be documented accordingly. In these cases, the construction auditor must regularly critically question how it is ensured that the procedure is exactly as declared by the parties involved.

With the help of the selection matrix from Table 3.2, it is possible to derive from the intersection of the audit area and audit field which documents come into question for the expert interviews. Based on the problem described in the initial situation and the objective – responsible clarification of quantity overruns – the required documents from Table 2.14 (cf. page 88) can be specified as follows:

- Contracts with planners and executors,
- Site Documentation,
- Construction Journal,
- Coordinator for safety and health matters protocols,
- Dimensions,
- Correspondence,
- Settlements, and
- Addenda.

The total contractor has also described processes, participants and responsibilities in detail in its project and organisation manual.

There you will also find, for example, a flow chart and a verbal description of how to deal with supplementary services – i.e. services that were added to the subcontractor's scope of services after the contract was concluded – and how the supplementary audit is to be carried out. It is precisely defined who has to check which documents and who is to be consulted in which cases for an opinion.

The findings from the literature study are, as already mentioned, verified and supplemented by expert interviews. In this context, experts are understood as the persons who are or were involved in the implementation of the construction project. Interviewing the persons becomes a challenge in cases where the persons have changed companies in the meantime and are therefore not available or only available to a limited extent.

Due to the risk of collusion, the true background of the construction audit is not initially communicated. Directly involved persons from the construction supervision, as well as the subcontractors, are initially requested to clarify documentation deficits.

During the audit, extensive reference is made to the actual execution of the construction work. It must be taken into account that the construction project is still in progress at the time of the audit. The audit is therefore to be provided with a key date as the status date. Assured findings therefore refer to the part of the work performed up to the key date. Conclusions on the overall scope represent a forecast, which must be clearly marked as such.

For a quick and efficient examination, the assessment of the answers or statements must first be checked for plausibility. If there are any indications of contradictions, evidence documenting the facts should be checked immediately. If the plausibility check does not reveal any contradictions, the detailed check can be carried out at a later date on a sample basis.

Scheduling
A total of 30 weeks are planned for the audit, which is divided into three phases. In the first 3 weeks, the start phase, the documents are reviewed and the quality of the documentation is examined on a sample basis. In the main phase, the next 25 weeks, audit reports are initially prepared for individual matters and supplemented by interviews. In the final phase, the last 2 weeks, the above-mentioned audit reports for individual issues are finalized and compiled into an overall report.

3.2.3   Implementation

The review of the documents shows that of the 43 subcontractors commissioned, six companies had generated a volume of supplements amounting to more than 50% of their main contract – without these services resulting in significant changes to the construction design or significantly extending the construction period. For this reason, the supplement management of the contracted construction supervisor became the focus of the audit.

The subject of the audit is a total of 50 supplements from six subcontractors with a volume of more than EUR 50,000. The audit reports are structured as follows:

I. Basics

   First of all, the audit report begins with a description of the facts, a reference to the object and performance and a tabular presentation of all relevant activities in the course of the preparation and negotiation of the addendum in the form of a brief chronology. Although this basic chapter is dispensable for the processor at the moment of the audit – especially if this is repeated for each audit report of an audit engagement – it facilitates the decision-making process for third parties who are to draw conclusions from it.

II. Formal examination

   The total contractor has contractually agreed with all subcontractors how supplementary offers must be structured and supported by attachments. This greatly facilitates the audit, as a formal check for completeness of the documents can be carried out along a clear guideline.

   The aim of the formal check is to establish whether the supplement can be checked at all. This audit step therefore only checks whether the documents have been submitted in the agreed form, completely and legibly.

   If the documents are not complete, the supplement will be rejected due to lack of verifiability. The subcontractor is requested to submit the missing documents within 14 calendar days. If the documents are not available after 14 calendar days, the supplementary offer shall be deemed withdrawn as agreed.

   Only when the documents are complete and legible, the examination of the basis of the claim takes place.

   As part of the audit review, it is first necessary to examine which documents the subcontractor has submitted. Subsequently, the audit report of the construction supervision is examined. Finally, the results of the construction supervision are compared with a formal audit carried out by the construction auditor himself.

III. Examination on the merits

   (a) whether the information provided by the contractor is correctly presented,
   (b) the reason for the derogation,
   (c) the actual nature and extent of the deviations,
   (d) which other benefits, if any, are cancelled or reduced.

   According to the systematics of the VOB, the supplementary offer should be submitted before execution. This presupposes that the cause of the supplement is based on a change order by the client.

   In fact, in this case, a large portion of the addenda were not formulated until after execution because some subcontractors performed what appeared to be required on the job site. Only in the course of determining the billable services did they realise that these were not covered by the main contract. These services then had to be subsequently requested by the total contractor in the form of supplementary quotations.

   However, if the alleged discrepancies do not have a cause attributable to the total contractor's area of responsibility, they will be rejected for lack of a basis for a claim.

The subcontractor is requested to present and prove the missing basis for the claim within 14 calendar days. If the relevant documents are not available after 14 calendar days, the supplementary offer shall be deemed withdrawn as agreed.

Only when the basis for the claim is available is the amount of the claim examined.

Within the scope of the audit, the subcontractor's justification of the claim must first be examined. This is also compared with the justification in the audit report of the construction supervision. On this basis, the construction auditor himself assesses the basis of the claim and checks the claim on the merits position by position.

IV. Verification of the amount

Within the scope of the examination of the amount of the claim, the claim is examined with regard to its amount. At this point of the examination it is therefore no longer in question whether a claim exists at all. Rather, it must be clarified how high the subcontractor's claim is and whether it was calculated as agreed.

The amount of the claim in this unit price contract is determined by two factors: the quantity front rates and the unit prices.

Set Forwards

Within the scope of the construction audit, the supplementary quantities are to be examined as determined by the subcontractor in his tender. According to the systematics of the VOB, the quantities and masses provided are to be determined from drawings. If the performed quantities and masses do not correspond to the drawings, they have to be invoiced by allowance. This means that, theoretically, every execution would have to be preceded by a quotation based on a modified plan. This planning would be the basis for billing after execution.

In fact, in this case, a large portion of the addenda were not formulated until after execution because some subcontractors performed what appeared to be required on the job site. Only in the course of determining the billable services did they realise that these were not covered by the main contract. These services then had to be subsequently requested by the total contractor in the form of supplementary quotations.

Within the scope of the audit review, the subcontractor's alleged quantity overruns are to be checked first. These are also compared with the quantity determination in the audit report of the construction supervision. On this basis, the construction auditor himself makes an assessment of the quantities presented and checks them item by item.

Unit Prices

Pricing must be carried out in accordance with the systematics of the VOB on the basis of the agreed prices. The VOB differentiates between additional and changed services without making an exact distinction that allows a clear and unambiguous allocation. In many cases, it is not important to the contractors whether they receive payment for additional services or for changed services – what is important to them is that payment is made. For this reason, contractors usually avoid the exact differentiation and show "remuneration according to §2 VOB/B" as the basis of calculation.

The determination of the prices for the new services in §§2 (5) and 2 (6) VOB/B also does not differ significantly, so that from the point of view of pricing this distinction seems to be dispensable in many cases.

From the point of view of the construction audit, it is important that the price components in which a deviation from the agreed service is claimed are presented in a comprehensible manner. Price components that do not change must be accepted in accordance with the calculation. There is a particular potential for dispute in the price components that are not calculated on the basis of performance, but as allocations or surcharges for overheads.

Within the scope of the audit review, the unit prices offered by the subcontractor are to be checked first. These are also compared with the price determination in the audit report of the construction supervision. On this basis, the construction auditor himself makes an assessment of the quantities presented and checks these item by item.

V. Result of the audit

The results of the audit are summarised once again in key words. This allows the reader to see the main audit results at a glance and to follow up on the details of the audit or the notes (see below).

VI. Notes

Inconsistencies, gaps and/or contradictions are initially indications of a lack of compliance with the rules. Criminal, civil or labour law consequences cannot be derived directly from this. However, if there are indications that intentional or grossly negligent action has been taken, the findings of the construction audit, formulated as indications, can provide important clues for further investigations.

Such references may relate to the form and scope of documentation, to missing evidence and assertions without supporting documents, to an incorrect price check or even to legal issues, the follow-up of which should be carried out by specialist lawyers.

The 50 audit reports on the individual supplements form the overall audit report in the summary.

3.2.4   Audit Result

The audit came to the conclusion that of the total of 50 addenda audited, not a single addendum was audited as specified by the total contractor. There was no evidence of a formal audit certifying the auditability of a supplement for any of the supplements.

If the purely formal aspects are excluded from consideration, there is nothing further to object to in the case of 11 supplements (i.e. more than 20% of the supplements examined): the supplements are based on an undisputed change order by the total contractor, the effects on pricing are presented in a comprehensible and plausible manner with reference to the pricing of the main contract. The actual performance of services has been comprehensibly documented by the subcontractors, and the invoicing has been carried out correctly.

Seven supplements (i.e. almost 15% of the supplements examined) should have been rejected on the merits because the services actually provided did not represent a deviation from the contractually agreed scope of services.

In the case of 32 supplements (i.e. almost 65% of the supplements audited), documents were missing from the factual documentation to such an extent that it was not even possible to prove from the files that the subcontractors were actually on the construction site during the period of execution in question in order to perform the services invoiced. Nevertheless, the construction supervisor had signed off on the billings. Since the interviews did not provide any plausible explanations for this either, the claim of the subcontractors cannot be confirmed from the point of view of the construction audit.

Two subcontractors filed for insolvency before the completion of the construction review. The general contractor is taking legal action against four other subcontractors and against the construction supervision.

3.3   Example Process Audit

3.3.1   Initial Situation

A residential construction company is to develop and implement more construction projects on a large scale in order to generate urgently needed living space. It has therefore set itself the goal of increasing its portfolio of rental apartments by 20% within 5 years. The company is concerned that the existing organization and its project management workflows may not be up to the task.

The company awards the construction work mostly as partial lots in unit price contracts to subcontractors with whom the VOB is agreed. In order to be able to cope with the personnel-intensive coordination work involved in the growing number of projects taking place in parallel, the company has commissioned general and total contractors. It is hoped that this will save time and money, as the coordination tasks are also contracted out in this way. In addition, contracts are structured as lump-sum price contracts in order to increase cost certainty.

However, the aforementioned hopes are not entirely fulfilled. The in-house coordination effort is reduced, which allows the company to implement a larger number of projects without additional personnel. But even with the lump-sum price contracts, the actual costs clearly exceed the budgeted costs.

An analysis of the invoiced construction projects shows that the volume of supplementary claims is not reduced by the agreement of lump-sum price contracts. The company therefore commissions the construction auditing department to review supplementary contract management with the aim of identifying potential for improvement and recommending measures for improvement.

3.3.2   Preparation

Thus, a prophylactic audit is to be carried out. In Fig. 3.2, the audit is classified in the process overview for the preparation of the construction audit and the procedure is followed according to the middle column.

Audit Objective
The housing company decides to carry out a process analysis for the new construction of rental apartments with the aim of identifying weak points and improving the processes of future construction projects through targeted measures.

The following earnings and control parameters are to be taken into account:

1. Regularity (legal requirements and compliance guidelines),
2. Process reliability (transparency and quality of processing),
3. Economy (efficiency and effectiveness of processing).

It is therefore not a question of clarifying wrongful conduct or fraudulent acts. This aspect is very important and must be clearly communicated to those involved in order to relieve them of the worry of being the subject of a criminal investigation and having to fear personal consequences.

Audit Area
For the selection of the audit area, it was first considered in which of the three audit areas the greatest influence on the existing processes could be exerted, taking into account the initial situation and objectives.

The audit area Execution is not considered, because in this audit area the supplements can arise, but no influence is exerted on the audit process.

The audit area of planning services could be considered, since the planner usually has to inspect supplements as part of the object supervision and is required to do this in a comprehensible form. However, experience shows that every planner has a different understanding of an appropriate audit process. In the case of the housing construction company with several construction projects running in parallel, this leads to the fact that different checking processes are used. As a result, the quality of the supplemental review varies widely. The audits are not comparable.

In addition, the residential construction company often works with total contractors. In these cases, the total contractor cannot check his own supplements. It is therefore the responsibility of the project management to check or at least approve supplementary services. According to the exclusion principle, only the project management audit area remains in which the greatest influence can be exerted on the supplement process.

From the initial situation and the objective it follows that the investigation should refer to the audit area of project management.

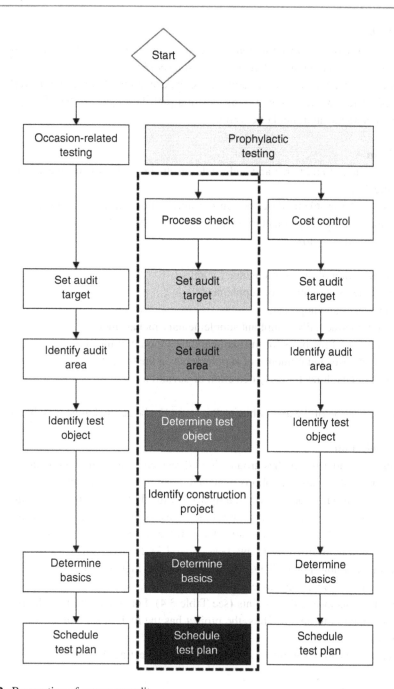

Fig. 3.2 Preparation of a process audit

Audit Field

In the present case, the examination is to take place retrospectively – i.e. ex-post. This leaves open in which audit field the investigation should start.

However, since the process of audit during the execution of construction work is to be improved, it follows from the starting position and the objective that the investigation should relate to the audit field of execution.

Audit Item

The intersection of audit area and audit field results in a selection of relevant audit items (see Table 3.2).

Based on the problems described in the initial situation and the objective – process improvement of the supplementary audit – the audit objects from Table 2.2 (cf. page 75) can be specified as follows:

- Documentation of the supplementary audit
- Monitoring of the volume of supplements
- Competences and approach
- Uniform contractually compliant supplementary management
- Examination of the basis for the claim and the amount of the claim
- Review of costs and influence of supplements on the budget
- Control of deadlines for early detection of deviations

After identification of the audit objects, the project to be audited must be selected.

Project Selection

The company under review has around 50,000 residential units in more than 250 properties in its portfolio. This portfolio is managed on an ongoing basis and is maintained and expanded through regular construction measures. In addition, 15 projects are in the planning stage and five major projects are under construction. A further two projects were being finalised at the time of the study. One of these two projects should be reviewed.

Project A was awarded 37 points (see Table 3.3). For two attributes the project was rated with maximum points. Thus, the project has the highest risk rating for 2 out of 10 attributes. Project B was also assessed in the same way.

Project B was awarded 32 points (see Table 3.4). For three attributes the project was rated with maximum points. Thus, the project has the highest risk rating for 3 out of 10 attributes.

Three aspects are taken into consideration when selecting the project:

1. Overall score.
2. Number of criteria in which a project received the highest possible rating.
3. Number of criteria in which a project received the highest score.

Table 3.3 Evaluation of a project A

Attribute type	Attribute characteristic	Points
Project size	More than EUR 100 million	5
Scope of change	More than 100 and not more than 500 supplements	4
Project complexity	Average complexity (e.g. large residential buildings)	3
Contract model	Global lump sum price contract	4
Budget compliance	Budget exceeded by max. 50	3
Deadline	Schedule exceeded by max. 25	2
Planning quality	More than 50 and not more than 100 notifications of concern	4
Completion of the planning	Implementation planning less than 40% complete	5
Fault scope	More than 50 and not more than 100 obstruction notices	3
Novelty	Project already realized in a comparable way	4
SUM		37

Project characteristics: Large multi-family house (has already been realised in a comparable way), EUR 120 million construction volume, 255 supplements, global lump sum price contract with functional performance description, budget exceeded by 45%, schedule exceeded by 20%, 73 objection notifications, execution planning 35% complete, 89 obstruction notifications

Table 3.4 Evaluation of a project B

Attribute type	Attribute characteristic	Points
Project size	More than EUR 100 million	5
Scope of change	More than 50 and not more than 100 supplements	3
Project complexity	Average complexity (e.g. large residential buildings)	3
Contract model	Global lump sum price contract	4
Budget compliance	Budget exceeded by max. 25	2
Deadline	Schedule exceeded by max. 25	2
Planning quality	More than 10 and not more than 25 notifications of concern	2
Completion of the planning	Implementation planning less than 40% complete	5
Fault scope	No more than 25 obstruction notices	1
Novelty	One-of-a-kind, the first of its kind	5
SUM		32

Project characteristics: Large multi-family building (has not been realized in a comparable way before), EUR 150 million construction volume, 98 supplements, global lump sum price contract with functional performance description, budget exceeded by 25%, schedule exceeded by 25%, 28 objection notifications, execution planning 30% complete, 17 obstruction notifications

The result is summarized in Fig. 3.3. In this figure, the evaluation of the attributes is shown as bars. The results for project A as black bars, for project B as white bars. Where the evaluation for project A and project B led to the same result, the bars are shown as the intersection of black and white – thus as grey bars.

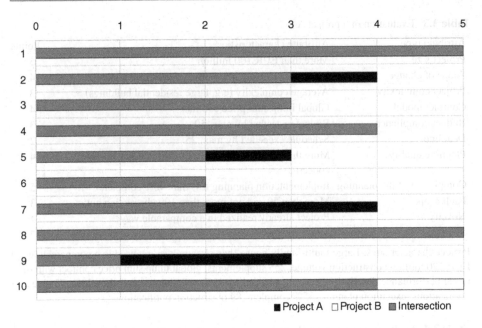

Fig. 3.3 Result of the project selection for the process audit

This type of display shows for which attributes a matching score was given (visible as grey bars), for which attributes project A received a higher score (visible as black bars) and for which attributes project B received a higher score (visible as white bars).

In the present case, project A was assessed with 37 points and project B with 32 points. For two attributes (Nos. 1 and 8), Project A was scored with maximum points, and for Project B, three attributes (Nos. 1, 8 and 10). For four attributes (Nos. 2, 5, 7, and 9), Project A was scored higher than Project B, and one attribute (No. 10) was scored higher for Project B than for Project A.

Project A therefore has the highest risk rating for 2 out of 10 criteria. Since Project A was also rated higher, with a score of 37, and thus as riskier, Project A was chosen for further investigation.

After the project has been selected, the bases required for the appraisal must be determined.

Basic Determination

The exam is mainly based on two types of sources. Firstly, the audit is conducted as a literature search. In this context, literature is understood to be all documents provided by the company. This occurs, for example, in the form of access to an electronic project folder or a web-based document management system.

On the other hand, the findings from the literature study are verified and supplemented by expert interviews. In this context, experts are persons who are involved in the execution of the project and who can contribute significantly to answering the question through their special knowledge.

With the help of the selection matrix from Table 3.2, it is possible to derive from the intersection of the audit area and the audit field which of the parties involved should be considered for the expert interviews. Based on the problem described in the initial situation and the objective – process improvement of the supplementary audit – the participants from Table 2.13 (cf. page 86) can be specified as follows:

- Owner's Contract Manager,
- Project Manager of the builder,
- Owner's Legal Department,
- If applicable, project manager for the selected project.

It should be possible to substantiate the responses of the parties involved by means of supporting documents. However, they can also be expressed as a statement and must then be documented accordingly. In these cases, the construction auditor must regularly critically question how it is ensured that the procedure is exactly as declared by the parties involved.

With the help of the selection matrix from Table 3.2, it is possible to derive from the intersection of the audit area and audit field which documents come into question for the expert interviews. Based on the problem described in the initial situation and the objective – process improvement of the supplementary audit – the required documents from Table 2.14 (cf. page 88) can be specified as follows:

- Global lump sum contract,
- Planning Contract,
- Procedural instruction supplementary audit,
- Documentation of existing supplements,
- Housing Authority Competency Matrix,
- Organizational charts of selection project and housing developer.

The examination may refer to the documents available in the selection project, but not yet to the actual implementation of subsequent projects, as these projects are still in the preparatory phase. All examination questions should therefore be formulated in the subjunctive or future tense.

Responses may likewise only express a desire or plan to address a task. Evidence that the same procedure has been followed is therefore not to be expected at this stage. The construction auditor must therefore regularly critically question how it is ensured that the same procedure is followed later as is planned now.

Scheduling

When carrying out an ex-post audit of the construction audit, it is important to quickly and quickly bring about a result to identify processes and to improve them. This task is carried

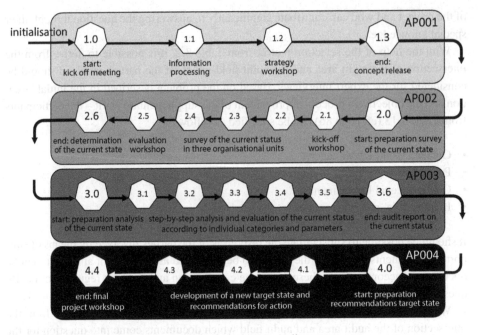

Fig. 3.4 Sequence of improvement of the supplementary audit process

out by the construction auditor in this example project within a period of 6 months. Figure 3.4 visualizes the process of the audit. The construction auditor divides the process analysis into four work packages.

Work package 01: Preparation.

The examination performance begins with a coordination meeting. This will essentially involve the presentation of the project team and the coordination and detailing of the schedule and the audit concept. Furthermore, the documents required by the construction auditor are agreed. In addition, the meeting offers the client the opportunity to focus on the analysis of the current situation and the required target processes.

In this phase, the audit basis is first compiled and the client's previous process culture is analysed. On this basis, the construction auditor compiles the audit questions and checklists and develops an audit plan. This specifies what is to be the subject of the audit in concrete terms – but also what is not to be included and to which aspects or which depth of investigation the construction auditor will restrict himself, if necessary.

Finally, the construction auditor schedules audit measures and interviews in coordination with the client and determines the process of document collection. The phase ends with a final meeting.

Work package 02: Analysis of the current situation.

This task block represents the beginning of the actual analysis work. The analysis is essentially divided into two steps.

In a first step, the current situation of the supplement management is recorded. This includes recording the project processes, determining roles and functions as well as competencies, recording the interfaces and communication channels. This is realized by analyzing the process management, by evaluating the handed over documents such as procedural instructions, organizational charts, audited supplements, contract documents, etc. and by targeted interviews.

This is followed by an interim reconciliation. The main purpose of this is to ensure that the construction auditor has correctly recorded all essential ACTUAL processes before moving on to the second step.

In a second step, the construction auditor draws up the process list and identifies the essential processes to be further developed.

The following detailed activities are to be calculated:

- Process analysis of the actual process with presentation of the usual participants, information paths, durations, times, tools used, etc.
- Analysis of processes of valid/applicable regulations (internal procedural instructions, organisation manual, regulations in HOAI, VOB, BGB etc.)
- Representation of the processes in suitable process diagrams
- Involvement of third parties (e.g. legal department, IT department) for the determination of process management requirements
- TARGET-ACTUAL comparison of the processes incl. evaluation
- On the basis of the findings from the above-mentioned analyses – compilation of a catalogue of criteria for essential processes to be further developed

Work package 03: Assessment of the current situation.

This task block builds on the analysis described above and is closely interwoven with it in terms of time and content. The evaluation is carried out with regard to compliance with the legal and operational requirements for economic efficiency, process reliability, throughput times of the processes, use of resources and adherence to the compliance requirements. Afterwards, the construction auditor consults with the client on the findings and evaluations obtained.

This phase ends with a summary of the current situation in an interim report as a presentation and the identification of alternative actions to improve the processes.

Work package 04: Presentation and evaluation of action alternatives for optimisation to the TARGET development.

This task block includes the development of recommendations. These are to be measures for the optimisation of the supplement management with regard to organisational structure, structure, process development, responsibilities, interfaces and communication channels. This task is carried out in close cooperation with the client.

The following detailed activities result for the construction auditor:

- Weakness analysis, including identification of time reserves and time wasters in processes, identification of risky processes, identification of process gaps, identification of economic processes,
- Presentation of the weak point analysis and joint discussion
- Summary of results and definition of process improvements

These recommendations are then presented, jointly discussed and, if necessary, modified. Finally, the recommendations are summarized in the form of a audit report. Figure 3.5 shows a summary of the contents of the four work packages.

3.3.3 Implementation

In order to identify the existing and previously practiced audit process for supplements, the existing procedural instructions and audited supplements with all associated documents are first analyzed. In addition, various questions support the work of the auditor. These are questions which the construction auditor asks himself during the literature research or which he asks the participants in the expert interviews. In addition to the auditor's own experience, information from the literature can also be used to compile suitable questions.

The contents of the implementation are the work packages 02 and 03 presented in the scheduling (see Fig. 3.5).

For the presentation of this example, essential questions for three audit items are presented in this chapter. For each question, the background that the question has for this

Preliminary phase and phase 1	Phase 1 and Phase 2	Phase 2 and Phase 3	Phase 3 and Phase 4
Work package 001	**Work package 002**	**Work package 003**	**Work package 004**
Content	**Content**	**Content**	**Content**
Creation of a development concept for the optimization of the supplement management.	Analysis of the ACTUAL situation in the departments that deal with the processing of supplements.	Analysis of the ACTUAL situation in the departments that deal with the processing of supplements.	Elaboration of action-attempts and recommendations for the TARGET state development
Destination	**Destination**	**Destination**	**Destination**
Targeted project planning and progress monitoring.	Collection and recording of all necessary information.	Assessment & evaluation of the situation & features.	Presentation of measures and for the optimization of processes
Start date: Month 01	**Start date:** Month 03	**Start date:** Month 07	**Start date:** Month 09
End date: Month 02	**End date:** Month 06	**End date:** Month 08	**End date:** Month 10

Fig. 3.5 Process improvement work packages

project is explained. Subsequently, the essential audit results for each question are stated and evaluated.

1. **Documentation of the Supplementary Audit**
   - How is it ensured that any changes that occur are sufficiently justified and documented in a comprehensible manner?
   - How is it ensured that the additional expenses from supplements can be used for onward charging according to their origin?
   - Have any standardised forms used for documentation been checked for legally compliant formulations and contractual conformity?
   - How is it ensured that the result of the negotiation with correction entries is documented in the original supplementary calculation in a comprehensible way?
   - How is it ensured that the current and uniform forms are always used in the company to document changes?
   - Is there a uniform understanding of how supplements are to be formally issued and is this understanding contractually fixed?

Construction Audit took a random look at some addenda on this project. It is noted that contractor change notices and addenda are reviewed by the General Planner and brought forward for submittal as award recommendations. However, these recommendations are not well understood by outsiders in terms of their system and reason and amount.

Thus, the construction auditor notes that no distinction between additional and changed services can be understood. However, this is of importance as the conditions of entitlement are different.

Some supplements show considerable increases in quantity, which should be critically examined in a new construction project. Possibly existing deficits in the tender documents. In this case, the addendum would have to be documented in such a way that it can be subsequently charged to the planner concerned. There is a regulatory gap in the review process here.

The construction auditor also notes that although there are forms and checklists in an in-house software solution, these are not used consistently by either the project management or the external planner. This does not ensure uniform documentation.

It is also not comprehensible how the cost estimate submitted was checked by the planner. There are poorly legible marginal notes, but there are no check remarks that establish a reference to the main contract, the local customary practice or other comparable calculation bases. The present procedural instruction also does not provide any clear regulation in this respect.

For some supplements, there is no documentation as to why the quantity estimates selected by the contractor are correct. Here, in fact, it was simply "ticked off".

The complete results of the process analysis concerning the documentation were recorded by the construction auditor in an ACTUAL process diagram. Gaps in the regulations were identified in the process diagram.

2. **Competences and Approach**
   - How is it ensured that the qualifications of the auditors with regard to the evaluation of supplementary services are sufficient?
   - Is there an internal separation of functions between planning and construction supervision?
   - How is it ensured that the audit quality of external planners meets the internal quality requirements?
   - Is a 4-eyes principle provided for and how is its implementation anchored in the process?
   - What formal regulations exist with regard to the approval of supplements?
   - Do process diagrams and descriptions exist, which guarantee a contractually defined procedure of the supplementary management between client and contractor?
   - How is it ensured that the submitted documents are complete in a uniform understanding of the contracting authority?
   - Does a defined supplementary process also provide for the securing of claims of the client against his contractor (anticlaim management)?

To answer these questions, the construction auditor looked at the organizational structure, the job descriptions of the employees, specifications from the project development and the general planning contract. It is found that there is no general separation of functions between planning and construction supervision.

The contract allows the same office to take on both tasks. However, there is a risk that deficits from the planning are deliberately concealed by the supplementary audit due to material defects that arise later and that the elimination of defects is thus declared to be an additional service. In this way, planning deficiencies remain undiscovered. The client unknowingly "pays" for the elimination of the defect by releasing the corresponding supplement.

This deficit is also caused by the fact that the client only checks the result of the supplement and does not describe the process of finding the result. This way is therefore not lived uniformly and not sufficiently documented.

In the view of the construction auditor, the professional competence of the auditors is sufficiently ensured by the job descriptions of the employees or by the tenders of the planners.

Dolose actions are made more difficult as the 4-eyes principle is practiced both internally and externally. A temporary job rotation, which takes place at least once a year in the form of a holiday replacement, supports this.

During the examination of the internal documents, a process instruction was found, but its content does not correspond to a process instruction with corresponding process descriptions. Thus, no contractually fixed process of the supplement management is depicted. This is the reason why there is no uniform process for supplement management.

Although the contracting authority has some tools for recording, a review of the supplementary agreements in hand does not reveal any systematic use of these tools on a regular and consistent basis.

The procedural instructions do provide information on which documents must be enclosed with an addendum. However, this instruction should also be included by the project management in the contract of the general planner and the contracts of the companies in order to ensure that all parties involved have the same understanding of how an addendum should be structured and which documents should be enclosed with it.

Due to the lack of a systematic approach, there is no guarantee that non-verifiable and unjustified supplements are justifiably rejected, as there is also no guarantee that all necessary aspects are checked in a supplement. A systematic examination – first of all of the grounds and then of the amount – is not discernible. A legal classification of the bases for claims – for example according to the exclusion principle – is not carried out.

A useful tool could be a checklist, which ensures that all necessary information for the supplementary audit is available. However, there is no such checklist or similar.

The complete results of the process analysis concerning the approach were recorded by the construction auditor in an ACTUAL process diagram. Regulatory gaps and risks were identified in the process diagram.

3. **Examination of the Basis of the Claim and the Amount of the Claim**
   - How is it ensured that the justifications for the changes and supplementary services are accurate and auditable?
   - How is it ensured that existing processes for different construction projects are suitable for different contract constellations.
   - How is it ensured that the supplementary audit process is described comprehensively and in a way that can be understood by all those involved in the audit?
   - How is it ensured that the supplementary audit process is carried out properly?
   - How is it ensured that the supplementary prices are checked properly and in accordance with the contractual agreements (market prices, original costing, actual costs, price databases, comparative projects, etc.)?

As previously noted, there is no review process for supplement management. Thus, it is not ensured that the amount of a supplement is not already checked before it is determined whether a claim exists at all. As long as the causality giving rise to liability has not been established, there is no need to check the amount of the supplement. A systematic approach to these steps thus leads to a more efficient working method and a regulated review process.

If these individual steps of the audit process are also given a time frame, the entire audit duration can even be defined in advance. This creates commitment and a certain calculability among all participants. In addition, this approach often supports a structured way of working.

The construction auditor has analysed the quality of the verification of the causality giving rise to liability on the basis of various supplements. It was found that the quality of the review varies considerably. It is not apparent that each auditor consistently adheres to systematically reviewing the bases of claims in relation to the subject matter of the contract. This does not exclude the possibility that claims bases are incorrectly assessed.

With regard to the examination of the liability-filling causality, the construction auditor has found that frequently costing attachments for the supplementary prices are not handed over with the supplement, which precludes an examination of the supplementary prices.

The complete results of the process analysis concerning the quality and sequence of the audit were supplemented by the construction auditor in the actual process diagram. Regulatory gaps and risks were identified in the process diagram.

3.3.4 Audit Result

In this chapter, the necessary measures and recommendations based on the findings are presented. The content of the implementation is the work package 04 presented in the scheduling (see Fig. 3.5).

The summary in the form of a so-called catalogue of measures is popular. In the following, the audit result for the example project is presented in this form. For audit items and questions for which it was determined that no measures are required (**level 1 – no significant risk – no measures required**), it is recommended that their audit results are not addressed in such a summary in order to keep the catalogue of measures clear.

It is also recommended that the audit results be compiled in tabular form with an assignment of responsibility and the naming of a target date.

Level 2 Low Risk: Recommend Appropriate Action
See Table 3.5.

Table 3.5 Catalogue of measures: recommendations due to low risk

No	Description	Responsible	Date
1	Construction auditing recommends using your own forms and not those of the contractor. This will speed up the audit process. At best, contractors are obliged to use this form		
2	The construction audit recommends the introduction of a management summary so that third parties can quickly identify the deviations and their causes		
3	The construction audit recommends carrying out a plausibility check in order to obtain an independent audit result and to check the work of the planner		
4	Construction auditing recommends developing a form as a formal template so that contractors always service a supplement using the same template. This ensures more complete addenda, creates efficiency and ensures equal quality in the addendum review		
5	The construction audit recommends to make a regulation on the validity of calculation/validity of prices		

Table 3.6 Catalogue of measures: recommendations due to high risk

No	Description	Responsible	Date
1	The construction audit urgently recommends the introduction of a change management system. It should be defined as a process whether/how/when a change in execution is to be documented and what follows from the change		
2	The construction auditing department urgently recommends defining a supplementary process that extends from the request for change, through the examination of the basis and amount of the claim, to the settlement of accounts. At the very least, a form is required here to standardise the documentation of the supplementary audit with standardised forms		
3	The construction auditing department strongly recommends that the negotiating strategy be determined and the negotiating scope defined prior to supplementary negotiations and, if necessary, that target agreements be reached with planners		
4	The construction review strongly recommends that the use of the negotiation protocol be specified and made part of the order. This could, for example, be a result sheet showing that the result has been agreed between the contractor and the client. This result sheet could be agreed as an annex to the supplement as a necessary document for the submission and review of the supplement		

Level 3 High Risk: Recommend Urgent Action

See Table 3.6.

In addition to the tabular summary of the measures, a recommendation in the form of a sample can also be handed over during a process audit by the auditor in order to facilitate the implementation of the measures. In the example project, a sample was handed over for the design of a structure for the supplementary offers, which in future should already be coordinated and agreed with the construction contract (cf. Fig. 3.6).

Such or similar specifications can be found in many project and organization manuals for construction projects. Such regulations mostly refer to material supplements. Addenda resulting from disruptions to the construction process cannot usually be included in such a scheme. So-called "construction time supplements" are therefore to be excluded from this with their own requirements for presentation and evidence.

It can remain undecided whether an audit of a supplement can actually be refused with reference to the above-mentioned regulation. It is probably indisputable that all parties involved benefit from the transparency of a well-structured process and from clear guidelines on the auditability of an addendum.

1. Cover letter
2. Annex 1: Claim cause
    a. Letter from the total contractor with order of performance
    b. Proof of the necessity of the service, if no order is available.
3. Annex 2 Reason for claim
    a. Designation of the basis for the claim (reference to VOB/B or contract)
    b. Presentation of the main contract performance (construction target)
    c. Presentation of the supplementary performance (deviation from the
       construction target))
4. Annex 3: Supplementary offer
    a. Formulated supplementary justification
    b. Spatial allocation of services
5. Annex 4: Dates
    a. Scheduling of the offered service
    b. Measures to prevent an extension of the construction period
6. Appendix 5: Costs
    a. Supplementary calculation
    b. Subsequent costing of subcontractor services
    c. Discontinued or reduced benefits
7. Annex 6: Miscellaneous
    a. Billing drawings (such as allowances, delivery bills, etc.)
    b. Supplementary LV

Fig. 3.6 Default for structuring an addendum

3.4   Example of Cost Control

3.4.1   Initial Situation

Due to urgently required modernisation measures and the expansion of medical care, a hospital company is exposed to extensive and very different construction measures. The hospital company finances these measures from its own funds and from external funds (subsidies).

In particular, the approved external funds must be drawn down by a fixed date and may not be exceeded. The budgets underlying the funding applied for correspond to the cost calculation from the planner's design planning. The hospital company has commissioned a general planner for the implementation of the various construction measures covering all service phases in accordance with HOAI.

The general planner is basically responsible for compliance with the upper limits set in the general planner contract. Nevertheless, the hospital association is concerned that these costs could be exceeded for various reasons. Possible overruns are to be identified quickly and acted upon at an early stage. The hospital association decides to have selected construction measures monitored by a construction audit.

3.4.2 Preparation

Thus, a prophylactic audit is to be carried out. In Fig. 3.7, the audit is classified in the process overview for the preparation of the construction audit and the procedure is followed according to the right-hand column.

Audit Objective
The hospital association's objective is to use an auditor who is independent of the project to monitor all costs as a matter of principle and to be alerted to any unlawful deviations. A further and more far-reaching objective is to detect cost deviations at an early stage, to identify their causes and to take appropriate measures to counteract further cost deviations. These can and may also be measures that have a cross-project effect.

Audit Area
For the selection of the audit area, it was first considered in which of the three audit areas the greatest influence on the costs incurred or on the avoidance of possible cost overruns could be exerted, taking into account the initial situation and the objectives.

Since a general planner has been commissioned for the planned construction measures, it can be stated that several responsible activities have been assigned to him through his contractual relationship, which can have an influence on the costs.

First of all, the general planner has contributed significantly to the determination of the calculated and provided total budget through his cost calculation from the design planning. If all construction services were awarded to different planners, he would not have to bear this responsibility alone.

Furthermore, the general planner is responsible for the respective implementation planning. The quality of the implementation planning is decisive for the quality and binding nature of the implementation. The higher the quality and binding nature of the implementation planning, the fewer adjustments and changes are to be expected during execution.

In addition, the general planner is responsible for the tender documents on which the award of the construction work is based. Clear and doubtless tender documents enable the executing companies to identify all required construction services and to take them into account in the bidding process. Due to fully developed tender documents, a contractor is able to calculate the construction work completely. Completely calculated construction services in turn lead to the fact that no or few unexpected changed or additional services have to be feared by the client afterwards.

And finally, the general planner supervises the execution of the construction work. He is thus responsible for the professional and defect-free production as well as the proper billing of the construction services.

It follows from the initial situation and the objectives that the investigation should relate to the area of planning services.

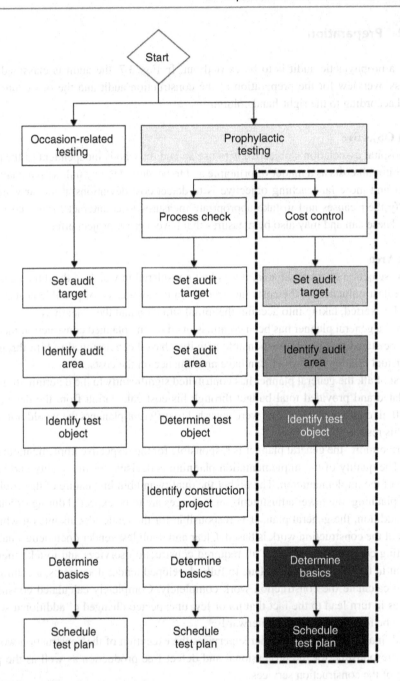

Fig. 3.7 Preparation of a cost control

Audit Field

In the present case, the audit is to take place during the project – i.e. ex-ante. This simplifies the identification and definition of the audit field. Since at the time of the audit primarily the execution of the construction work is pending, it follows from the initial situation and objective that the investigation should relate to the execution audit area.

Audit Item

The intersection of audit area and audit field results in a selection of relevant audit items (see Table 3.7).

Based on the problems described in the initial situation and the objective – control of costs – the audit objects from Table 2.2 can be specified as follows:

- Monitoring of the execution in compliance with the regulations and the tender,
- Control costs and release budgets based on performance,
- Control of deadlines for early detection of deviations,
- Control of the quality and timely demand of the evidence,
- Verification of the measurements in accordance with the rules,
- Supervision of required hourly wage work,
- Factual and mathematical verification of accounts,
- Review of anticipatory change management,
- Quality of review of supplements.

After identification of the audit objects, the project to be audited must be selected.

Table 3.7 Delimitation of audit area and audit field Example 3

	Test area: Project management	Test area: Planning services	Test area: construction services
Audit field: Project preparation			
Audit field: Planning			
Audit field: Preparation for execution			
Audit field: Execution			
Audit field: Project completion			

Project Selection

The hospital company is carrying out several different construction measures. The question therefore arises as to which project should be selected for monitoring by the construction audit. For this purpose, the project selection procedure already presented in Sect. 2.4 was applied.

Some projects are still in preparation and are therefore not suitable for investigation in the selected audit field of execution. Two projects were under implementation at the time of the study. One of these two projects is to be subjected to the review (Table 3.8).

Project C was assessed with 28 points. For one attribute, the project was rated with maximum points. Thus, the project has the highest risk rating for 1 out of 10 attributes. Project D was also rated in the same way (Table 3.9).

Project D was rated with 31 points. For three attributes the project was rated with maximum points. Thus, the project has the highest risk rating for 3 out of 10 attributes.

Three aspects are taken into consideration when selecting the project:

1. Overall score.
2. Number of criteria in which a project received the highest possible rating.
3. Number of criteria in which a project received the highest score.

Table 3.8 Evaluation of a project C

Attribute type	Attribute type	Attribute type
Project size	More than EUR 50 million and not more than EUR 100 million	4
Scope of change	More than 25 and not more than 50 supplements	2
Project complexity	Very high complexity (e.g. university hospitals)	5
Contract model	Unit price contract with individual award	1
Budget compliance	Budget respected or exceeded by max. 10%.	3
Deadline	Schedule exceeded by max. 25	2
Planning quality	More than 50 and not more than 100 notifications of concern	4
Completion of the planning	Final design at least 80% and less than 100% completed	2
Fault scope	No more than 25 obstruction notices	1
Novelty	Project already realized in a comparable way	4
SUM		28

Project features: Modernisation and reconstruction of the existing wards during ongoing operation, addition of a standard storey (already realised in a similar way), EUR 75 million construction volume, 35 supplements (at the start of the appraisal), individual awards with unit price contract incl. Specifications, within budget (at the start of the appraisal), schedule exceeded by 15%, 55 objection notifications, implementation planning 85% complete, 11 obstruction notifications

Table 3.9 Evaluation of a project D

Attribute type	Attribute characteristic	Points
Project size	More than EUR 100 million	5
Scope of change	More than 50 and not more than 100 supplements	3
Project complexity	Very high complexity (e.g. university hospitals)	5
Contract model	Detailed lump sum price contract	3
Budget compliance	Budget exceeded by max. 25	2
Deadline	Schedule adhered to or exceeded by max. 10%.	1
Planning quality	More than 10 and not more than 25 notifications of concern	2
Completion of the planning	Final design at least 40% and less than 60% completed	4
Fault scope	No more than 25 obstruction notices	1
Novelty	One-of-a-kind, the first of its kind	5
SUM		31

Project characteristics: Extension to laboratory building (has not yet been realised in a comparable way), EUR 105 million construction volume, 60 supplements, global lump sum price contract with functional performance description, budget exceeded by 25%, on schedule (at the time of the start of the audit), 25 objection notifications, execution planning 50% complete, 25 obstruction notifications

The result is summarized in Fig. 3.8. In this figure, the evaluation of the attributes is shown as bars. The results for project C as black bars, for project D as white bars. Where for project C and project D the evaluation led to the same result, the bars are shown as the intersection of black and white – thus as grey bars.

This type of presentation makes it visible for which attributes a matching score was given (visible as grey bars), for which attributes project C received a higher score (visible as black bars) and for which attributes project D received a higher score (visible as white bars).

In the present case, Project C was awarded 28 points and Project D was awarded 31 points. For one attribute (No. 3), Project C was scored with maximum points, while for Project D it is three attributes (Nos. 1, 3 and 10). For three attributes (Nos. 5, 6, and 7), Project C was scored higher than Project D. Five attributes (Nos. 1, 2, 4, 8, and 10) were scored higher for Project D than Project C.

Project D therefore has the highest risk rating for 3 out of 10 criteria. Since Project D was also rated higher overall, with a score of 31, and thus as riskier, Project D was chosen for further investigation.

After the project has been selected, the bases required for the appraisal must be determined.

Basic Determination

The audit is mainly based on two types of sources. First, the review is conducted as a literature search. In this context, literature is understood to be all documents provided by the

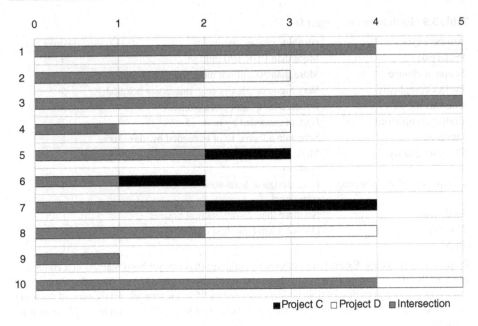

Fig. 3.8 Result of project selection for cost control

general planner. This takes place, for example, in the form of access to an electronic project folder or a web-based document management system.

On the other hand, the findings from the literature study are verified and supplemented by expert interviews. In this context, experts are persons who are involved in the execution of the project and who can contribute significantly to answering the question through their special knowledge.

The selection matrix in Table 3.7 shows, as an intersection of audit area and audit field, which of the participants could be considered for the expert interviews. Based on the problem described in the initial situation and the objective – control of costs – the participants from Table 2.13 (cf. page 86) can be specified as follows:

- Project manager of the general planner,
- Construction Supervising Planner, Auditor,
- General Planner's Contract Manager,
- Project Manager of the builder,
- Owner's Legal Department.

It should be possible to substantiate the responses of the parties involved by means of supporting documents. However, they can also be expressed as a statement and must then be documented accordingly. In these cases, the construction auditor must regularly critically question how it is ensured that the procedure is exactly as declared by the parties involved.

The selection matrix in Table 3.7 shows, as an intersection of audit area and audit field, which of the documents could be considered for the independent literature search.

Based on the problems described in the initial situation and the objective – control of costs – and the project selection, the documents from Table 2.14 (cf. page 88) can be concretised as follows:

- Audited performance levels,
- Audited accounts,
- Audit reports on supplements,
- Documentation of deadline tracking,
- Documentation Budget Tracking.

Scheduling

When carrying out an ex-ante construction audit, it is not a matter of bringing about a result in the short term and quickly in order to establish processes or to clarify certain irregularities. Rather, it is a matter of systematically monitoring a project over a certain period of time. In the example shown here, the construction audit accompanied the entire period of execution of the construction project, which was 24 months. Within the scheduling it was therefore determined to which events the construction audit is to be invited and in which regular or irregular events it is to be involved.

The audit begins with an introduction to the project and the main contracts. The construction auditors then obtain an initial overview of the current billing status and check the measurements, billing statuses, evidence for billing, additional cost notifications and resulting supplements as well as the audit documents for supplements that have already been submitted. These processes take about 4 weeks.

Subsequently, the first results are analysed, evaluated and submitted to the client with initial proposals for any necessary adjustments.

In a further step, based on the results of the first analysis, it is agreed with the construction auditor for which audit objects regular integration should be set up. In the example outlined here, the construction auditor was regularly integrated into the processing of the additional cost notifications, the supplementary audit and the settlement of construction work after the initial analysis.

The work results were documented once a month by preparing and handing over a short report and presented to the client.

3.4.3   Implementation

To identify cost overruns and their causes, the comparable planned and actual costs were first compared. This approach provides initial indications and allows an approach to the neuralgic issues.

In addition, various questions support the work of the auditor. These are questions that the construction auditor asks himself during the literature research or that he asks the participants in the expert interviews. In addition to the auditor's own experience, information from the literature can also be used to compile suitable questions.

For the presentation of this example, essential questions for three audit items are presented in this chapter. For each question, the background is explained, which the question has for this project. Subsequently, the essential audit results for each question are stated and evaluated.

1. Release and Review of Hourly Wage Work

The occurrence of hourly wage work is often unavoidable on construction sites. Construction projects are unique projects. In this respect, situations inevitably arise in which unforeseen small tasks arise that are not covered by any construction contract agreement. These are often activities that have to be completed at very short notice or for which it is not worth setting up, coordinating and agreeing a supplementary service.

In reviewing the project example with respect to hourly wage work incurred, questions include.

- Does the general designer have appropriate authority to agree to hourly rate work?
- How is it ensured that hourly wage work is not carried out at the same time as work from the main contract (risk of double billing)?
- Is it ensured that contractual regulations on hourly paid work are adhered to (right to issue instructions, form of documentation, written form, recognition, control, adherence to deadlines)?
- Is the requirement for hourly work by the general planner accurately justified?
- Prior to release of hourly work by the general planner, is it checked whether it can also be awarded as a supplementary agreement?

A review of the billing for drywall services shows that they have increased from an original amount of approximately EUR 1.4 million to approximately EUR 1.8 million. Closer examination of this increase has shown that hourly wage work has arisen continuously over the entire period of execution in addition to the main services.

As a result, the contents of the time sheets were examined and it was found that, as a rule, there is no concrete description of what additional work was carried out and why this was necessary. Rather, the impression is often given that additional expenses from the elimination of defective services have been settled. This is evidenced by entries such as this, for example:

- "Closure of breakthroughs not made according to plan".
- "Multiple opening and closing of suspended ceilings in stores and replacement of panels and repair of substructure."
- "Dismantling of assembled door and frame and storage in site office, dismantling plasterboard wall and disposal container".

The above points make it clear that the hourly wage work of the executing company is due to inadequate construction supervision and inadequate coordination of the trades. From the comparison of the construction diary with hourly wage slips it is not evident whether the personnel employed carried out work from the main services at the same time.

Furthermore, the evidence does not show where the activities took place. Thus, the use on site cannot be reconstructed. It is not possible to verify the additional work.

In addition, it was established that the planner had the work carried out despite the fact that the contracts expressly state that hourly wage work may only be carried out at the request of the client and must be submitted for approval on the day following the work.

Assuming that the services performed by hourly labor were actually necessary and unavoidable, the owner could not have avoided the increased costs. However, the question arises as to why these enormously extensive services were not requested from the contractor as supplementary services. After all, hourly wage work is by far the most expensive activity on a construction site. Additional services offered as supplements offer the possibility of negotiation and thus of achieving an economic result. In addition, the scope of services is clearly formulated in a supplement and the success of the work is defined. After completion of the supplementary work, the result is therefore measurable.

The construction audit has thus come to the conclusion that the client has incurred excessive costs for the execution of hourly wage work. The occurrence of this is not comprehensible. Measures must be taken to avoid or reduce the incidence of further hourly wage work.

2. **Audit of Accounts**

The auditing of invoices is a service obligation of the general planner commissioned with the supervision of the project. When reviewing the project example with regard to the release of partial and final invoices, the following questions are asked, among others:

- Are the services invoiced contractually agreed?
- How is it ensured that only services actually provided are invoiced?
- Do quantities agree with the measurements and prices with the contract?
- Are there significant discrepancies between the tender and the settlement?
- How is it ensured that there is no double billing in the main order and supplements?
- Are discounts, rebates, etc. taken into account in accordance with the contract?
- Is the accounting mathematically correct?
- Budget compliance – is there possibly an overpayment?
- How is monitoring of due dates carried out?
- How is it ensured that the client is made aware of invoice reductions in the event of defects?
- Are forecasts of cost developments made in a timely manner?
- How is it ensured that the use of personnel, materials and construction equipment is comprehensively and verifiably documented with a view to later proper accounting?

- How is it ensured that invoicing is carried out in a comprehensible manner in accordance with the rules of the VOB/C, i.e. the DIN standards (if agreed)?

After the audit, the general planner shall forward his audit result to the client with corresponding recommendations. The general planner's audit opinion does not constitute an acknowledgement of the contractor. If the client pays too much due to an incorrect audit, the general planner may be held liable.

In the present example, the construction auditor entered the audit at an early stage of execution. Some of the questions mentioned can therefore not yet be fully answered.

For example, the question of whether there are significant discrepancies between the tender and settlement or for certain items can usually only be answered well at the end of the project. Unless the structure of the tender and settlement is subject to the project structure of the construction project (breakdown by components, levels, rooms, etc.). In this case, deviations between tender and settlement can be measured at any time.

The example presented is a detailed lump-sum contract. This means that the agreed services are recorded in detail in a bill of quantities and calculated by the company, but the billing of the services is based on a lump sum. In this case, the quantity risk is borne by the contractor.

The flat rates can be invoiced in a payment plan or as partial flat rates for individual service packages – as in the example project. A down payment on individual partial lump sums is also conceivable. In this respect, despite the lump sum, the question arises as to how the performance status on the construction site is measured so that a fair settlement of the partial lump sums is possible?

By checking these prerequisites on the example project, the construction auditor determines that the agreed detailed and lump-sum bill of quantities cannot be transferred to the project structure without further ado. The basis of the settlement is therefore a percentage estimate for each partial lump sum by the general planner.

The percentage estimate is made on the basis of the general schedule. The general planner's estimate does not include any other documents that provide evidence of the services actually performed on the construction site. In this respect, overpayment or underpayment by the contractor cannot be ruled out.

Furthermore, the construction auditor checked the invoice amounts and found that the general planner was unable to deduct the contractually agreed discounts on several occasions due to delayed checks of partial invoices. In total, the unrealized discounts amount to approximately EUR 70,000.

The analysis of the related correspondence showed that in some cases the client repeatedly drew attention to the fact that the agreed discount period had repeatedly been exceeded due to the delayed processing of invoices.

The construction audit has thus come to the conclusion that the client incurred excessive costs in the settlement. These are due to a delay in the auditing of the accounts. Measures must be taken to avoid delaying further audits.

In addition, it cannot be ensured that the billing status always corresponds to the construction status, as the invoice verification is insufficiently documented. Measures must be taken to make the audit more transparent.

3. Release and Review of Supplements

The creation, preparation, review and approval of a supplementary performance generally raises a whole series of questions for a construction auditor.

If services have changed compared to the original contract, these changes must be identified, evaluated and reported in accordance with the contract. The claim resulting from this notification must be presented by the contractor on the merits and subsequently examined by the general planner. This is called the claim basis review.

The claim basis is a necessary prerequisite for further processing and the calculation of the claim amount. If a supplement is justified on the merits, the claim amount is checked.

Questions asked during the review of the project regarding supplemental processing included the following.

- Are the justifications for the changes accurate and verifiable?
- Are the prices of the individual items broken down according to the individual costs of the partial services for comprehensible pricing?
- Are the prices reasonable and do they correspond to the calculation basis?
- Have the supplementary offers been checked for their cost-effectiveness in a comprehensible manner?
- How is it ensured that there are no overlaps between supplementary claims and services from the main contract?
- When checking the supplementary services, is attention paid to any necessary compensation calculation of overheads?
- How is it ensured that the audit results are comprehensible by a third person?

For the example project, about 60 supplements had arisen by the time the construction audit was started. Since the project is an extension of the existing clinic complex and thus a new building, the question arises as to the reasons for the additional services required. After all, a general planner has the opportunity to include all necessary services in a newly planned building.

For this reason, the construction auditor first analyzed and sorted the supplements according to the amount of the costs. After all, the existence of an addendum says nothing about whether these are actually changes that are associated with increased costs. It is just as likely that there was only a replacement of materials, which has a relatively cost-neutral effect on the budget.

The analysis of the costs per supplement identifies which supplements have contributed to significant cost changes. According to this, the structural work in particular has increased by around 25% due to various supplements. A closer examination of the circumstances that led to this increase shows that the roof construction of the extension was changed after completion of the execution planning in order to optionally create the possibility for an additional landing pad for the rescue helicopter.

The further in-depth review of these facts leads to the conclusion that this is a subsequent change request by the client. This is due to a lack of coordination with the decision-makers of the hospital company at the time of the project development.

Based on this example, the question arises as to how the client can ensure that potential risks for the occurrence of supplements are identified at an early stage. This is important in order to avoid supplements if necessary or to initiate budget increases in good time in order to avert possible construction delays – and thus further costs – due to a lack of financial viability in the meantime.

This is often referred to as anticipatory claim management. This includes the conscious checking of the tender documents for possible ambiguously formulated services. This is a measure that can only be taken in advance. During execution, this problem can no longer be cured.

However, foresighted claim management can also be described as a vigilant response by the client to the contractor's concerns. These are often indicators of inadequate planning services and harbingers of later cost increases.

Furthermore, during the execution, the follow-up of obstruction notices as well as the examination of their effects on the construction process or the execution are part of the possibilities of a foresighted claim management.

In addition, the audit quality of the supplements is checked by the construction audit. This shows that there is no comprehensible justification for the audit result. Various supplementary items are only "ticked off" by the responsible auditor of the general planner. Thus, there is a risk that the agreed prices are too high. If such a procedure is approved by the client, there is also the risk that the general planner will take advantage of this circumstance to carry out fraudulent acts.

The construction audit thus came to the conclusion that the client incurred excessive costs due to supplements. These are due to change requests of the client after completion of the execution planning. Measures must be taken to avoid future changes during execution or to reconsider their effects before initiating them.

3.4.4 Audit Result

In this section, the necessary measures and recommendations based on the findings are presented. The summary in the form of a so-called catalogue of measures is popular. In the following, the audit result for the example project is presented in this form. For audit items and questions for which it was determined that no measures are required (**level 1 – no significant risk – no measures required**), it is recommended that their audit results are not discussed in such a summary in order to keep the catalogue of measures clear.

It is also recommended that the audit results be compiled in tabular form with an assignment of responsibility and the naming of a target date.

Level 2 Low Risk: Recommend Appropriate Action
See Table 3.10.

Level 3 High Risk: Recommend Urgent Action
See Table 3.11.

Table 3.10 Catalogue of measures: recommendations due to low risk

No	Recommendation of appropriate measures	Responsible	Date
1	It is recommended that the determination that the general designer does not have authority to agree to hourly labor be more closely followed and sanctioned as appropriate		
2	It is recommended to carry out stricter controls – Also on the part of the contractor – and to punish him if necessary, so that he does not allow the planner to release any management reports in the future		
3	It is recommended that the general planner be made more responsible for processing the accounts promptly in order to be able to deduct the agreed discounts, if necessary with the agreement of sanctions or the establishment of a suitable control system		
4	It is recommended to establish a forward-looking claim management and to install it by the general planner or an external auditor		

Table 3.11 Catalogue of measures: recommendations due to high risk

No	Recommend urgent action	Responsible	Date
1	It is recommended that the handover of hourly wage statements be regulated more precisely so that they are available promptly. In this way, a better control of the number and quantity of required hourly wage work can be achieved and, if necessary, a conversion into a more calculable supplementary service – As far as possible – Can be made		
2	It is recommended that a specification be made as to the (uniform) form in which hourly wage work is to be documented in order to ensure that work from the main order is not carried out at the same time		
3	It is recommended to introduce a comprehensible and verifiable performance determination, which forms the basis for the settlement of the partial lump sums. The settlement of realistic partial lump sums forms the basis for the detection of significant deviations between tender and settlement already during execution. In addition, it provides reliable forecasts of the cost development		
4	It is strongly recommended to improve and standardize the quality of the supplementary audit. The audit results must be comprehensible by a third person. The audit on the merits must be unambiguous and refer to the contractual agreements		

Evaluation of the Construction Audit

4

4.1   Introduction

The results of the audit are presented in a report, the appropriate form of which depends on the circumstances of the individual case. It can, for example, take the form of an audit report, an expert opinion on the subject matter of the audit or a workshop.

The form, content and scope of the presentation of results varies depending on the subject of the audit and the objectives pursued by the client of a construction audit. If the client has not already formulated this in the invitation to tender for the construction audit service, it is recommended that this be agreed in advance of the inspection.

4.2   Audit Report

Binding regulations for the preparation of an audit report that presents the results of the construction audit do not currently exist in Germany. However, the IIA standards of the 2400 Group provide helpful guidance as they address reporting.

Requirements for internal audit reporting are formulated primarily in IIA Standards 2410 and 2420, and are further substantiated by implementation standards and implementation guidelines formulated in the form of practical advice.

For example, IIA Standard "2410 – Reporting Criteria" requires that reporting include objectives and scope and results of the engagement. The standard also recommends that satisfactory performance be recognized in the reporting.

The results can be presented in the form of conclusions, recommendations and action plans. For this purpose, they can be built up as a causal chain, as shown below by way of example:

© The Author(s), under exclusive license to Springer Fachmedien Wiesbaden   139
GmbH, part of Springer Nature 2023
P. Wotschke, G. Kindermann, *Construction Auditing*,
https://doi.org/10.1007/978-3-658-38841-6_4

1. Actual and target situation are compared. The deviations between the actual and target situation are initially determined in a value-free manner by means of relevant **factual descriptions**. Causes for the deviation and the resulting risks or effects are explained if they are known and can be presented objectively.
2. From the findings (point 1), **judgements** about their implications and **conclusions are** derived. Conclusions can take into account sub-areas as well as the entire audit engagement and must be clearly declared as such. These in fact represent the (subjective) assessment of the auditor.
3. In the form of **recommendations**, generally or specifically designed options for action can be identified that may be suitable for improving the assessment or the conclusions (point 2). In many cases, concrete corrective measures are recommended.
4. For the implementation of recommendations (point 3), **action plans** are drawn up, which contain specifications on the timing and content, as well as naming participants and responsibilities. The content of the action plan can contain both general measures and concrete proposals for improvement.

IIA Standard "2420 – Quality of Reporting" defines the quality of reporting and requires audit reports to be prepared in a timely manner and to have the following characteristics:

- Correct, that is, free of errors and distortions.
- Objective, i.e. factual, impartial, unbiased and balanced assessment.
- Clear, so easy to understand and logical, without unnecessary technical terms.
- Concise, that is, without unnecessary explanations, superfluous details, double statements and long-windedness.
- Constructive, i.e. supporting the client as well as the organization and thus leading to the necessary improvements.
- Complete, i.e. include all essential and relevant information and findings to explain the recommendations and conclusions (Table 4.1).

4.3   Expert Opinion on the Test Object

In cases where a construction audit is carried out on an ad hoc basis, a brief audit report is often not sufficient to present and justify the results. If irregularities are identified, they must be thoroughly questioned and explained. This is particularly necessary if there is a risk of criminal and/or civil law consequences.

It is therefore of particular importance that statements of fact are formulated precisely. Assessments and conclusions derived from facts must be identified as such and must be equally precise. Statements by third parties must be accompanied by a reference to the source, and conclusions drawn from statements by third parties must be verified wherever possible. If verification is not possible, this must be indicated. If contradictions are found in statements or conclusions, these must be named and discussed.

Table 4.1 Group of 2400 standards

IIA standard	Implementation guideline	Implementation standard
2400 reporting	2400-1: legal considerations in the dissemination of results	
2410 reporting criteria	2410-1: reporting criteria	2410.A1 final communication of examination results
		2410.A2 recognition of satisfactory performance
		2410.A3 publication of results to parties external to the organisation
		2410.C1 communication of results in consultancy assignments
2420 quality of reporting	2420-1: quality of reporting	
2421 errors and omissions		
2430 use of wording performed in accordance with the international standards for the professional practice of internal auditing		
2431 disclosure of non-compliance with standards under the contract		
2440 dissemination of results	2440-1: dissemination of results 2440-2: dissemination of confidential information inside and outside the reporting channel	2440.A1 responsibility of the internal Audit manager for communicating results
		2440.A2 assessment of the conditions for publishing results outside the organization
		2440.C1 responsibility of the head of internal audit for communicating results of consulting engagements
		2440.C2 communicate significant findings identified during consulting engagements

The written report should be structured according to the technical reports of construction experts with the following structure.[1]

---

[1] Keldungs, K.-H., Ganschow, J., Arbeiter, N. (2018): Leitfaden für Bausachverständige. Legal bases – expert opinions – liability. Springer Vieweg, Wiesbaden p. 24ff.

- Table of Contents
- Initiation
- Rendering of the task
- Procedure
- Site visits/Interviews
- Documents used
- Description of facts
- Assessment and conclusions
- Summary
- Attachments

4.4   Workshop

The results of appeal proceedings are in fact in the vast majority of cases not the basis for criminal proceedings. Nor is it in keeping with the self-image of auditing to be seen as a tool of the criminal prosecution authorities. On the contrary, the aim is to promote behaviour that complies with the rules and to improve unsafe, sometimes even ignorant behaviour in order to prevent misconduct.

Thus, the communication of the results is not only about negative aspects, but also about positive aspects that have come to light during the review.

Particularly in problematic situations, regulatory gaps and cases of conflict, positive and negative findings are collected in construction projects. If this knowledge, similar to a "lessons learned", can also be used for other cases, the probability of future project successes is increased.

"Lessons Learned" is a method to systematically collect experiences (positive as well as negative) made in the project and to draw insights from them in order to improve the handling of future plans and projects. The method is now a common component in the project management method toolbox.[2]

The findings, the lessons learned, are often communicated in workshops, as in the case of the construction revision. It is important to work out the lessons learned together with the relevant project participants. This can be done either by means of workshops or individual interviews. This should take place at intervals of a few weeks, as the later development has the advantage that those involved see the facts with a certain distance and thus less emotionally.

In the context of the workshop, as already mentioned, both the findings that are free of objections and those that give rise to criticism should be communicated.

---

[2] Adler, N. (2015): Tools for Project Management, Workshops and Consulting: Compendium of the most important techniques and methods. 6th revised edition. Publicis.

Once these two groups of cases have been dealt with, concrete tips, suggestions for improvement or recommendations are developed for the points that did not run optimally. Since these often cannot all be implemented immediately, it is advisable to develop an action plan. The participants can be involved in this process, which leads to a better acceptance of the measures and possibly also to ideas that are developed individually from among the participants. Working in a group is very often a creative and collaborative process that can lead to high quality results.

However, a good result also depends to a large extent on good preparation of the workshop. The better the auditor discusses the procedure with the client and mentally runs through the course of the workshop, the higher the chances for a successful implementation.

In doing so, it is important to consider which specific issues are to be addressed and which goals are to be achieved. It has also proven useful not to have a workshop conducted and led by one person alone. A distribution of tasks not only relieves the burden, it also enables work with distributed roles, so that different positions and perspectives can be taken without this appearing discordant to the participants. This is where external auditors can make a valuable contribution, as they can make findings from the outside that would not be accepted by internal auditors.

The implementation of the workshop can follow the following structure:

In order to ensure that the workshop is remembered by the participants and thus has a lasting effect, it is advisable to follow up the workshop. This includes, for example, a photo documentation of the workshop results, which is sent to the participants. The results should also be passed on to the client or the responsible functionaries in the company (Table 4.2).

Table 4.2 Structuring example of a workshop

TOP	Title	Description
1	Opening	Welcome, introduction, agenda
2	Objective	Explain the concrete objectives of the construction audit, methods and procedures
3	Game rules	Documentation of results, communication rules
4	Facts	Positive and negative findings
5	Conclusions	Evaluation of the findings
6	Recommendations	Elaboration of approaches for improvement
7	Measures	Options for implementing the recommendations
8	Feedback	Integrate participants, confirm consensus
9	Further procedure	Assignment of responsibilities and deadlines

Final Thought

5

With this book, we set ourselves the task of explaining the exciting, diverse and varied field of activity in construction auditing and thereby improving the image of construction auditing among those involved in construction.

Since construction auditing is an interface discipline, we have briefly discussed the basics of auditing as well as construction management.

Our experience from recent years has shown that it is often very difficult for the initiators of a construction audit to make the right decisions on their own when selecting the right inspection fields and inspection objects or when choosing the right project. Even the formulation of the objectives is not always easy. It is not uncommon for all those involved to find out what the right objective would have been only on the basis of the results of a construction audit.

Likewise, it has often been difficult for us as construction auditors to always give the right advice to the very different companies. We have therefore developed the systematic approach that we have presented in this book. This systematic approach is by no means a dogma, but at best a guideline. It is intended to provide suggestions that can be used for the systematic preparation and execution of a construction audit.

With concrete – but anonymised – case studies, we have provided an insight into the preparation and execution of the construction audit and the application of our system. We have selected three different audit tasks to illustrate the broad spectrum of construction auditing services. For a comprehensive picture, one could write a separate book of examples from construction auditing.

Finally, we outlined how the results and findings can be evaluated and communicated.

If, with this book, we can have the effect that one or the other sees the construction auditing as an opportunity to improve himself and his work – not as a risk of having mistakes pointed out – then we will have come a good deal closer to the goal we are striving for.

© The Author(s), under exclusive license to Springer Fachmedien Wiesbaden
GmbH, part of Springer Nature 2023
P. Wotschke, G. Kindermann, *Construction Auditing*,
https://doi.org/10.1007/978-3-658-38841-6_5

References

Berwanger, Jörg; Kullmann, Stefan (2012): Interne Revision. Funktion, Rechtsgrundlagen und Compliance. 2. Aufl. 2012. Wiesbaden: Springer Gabler.

Brauweiler, Hans-Christian (2019): Risikomanagement in Unternehmen. Ein grundlegender Überblick für die Management-Praxis. 2., erweiterte und ergänzte Auflage. Wiesbaden: Springer Gabler (Essentials).

Brüggemann, Holger; Bremer, Peik (2020): Grundlagen Qualitätsmanagement. Von den Werkzeugen über Methoden zum TQM. Wiesbaden: Springer Vieweg.

Bünis, Michael; Gossens, Thomas (2016): Das 1x1 der Internen Revision. Bausteine eines erfolgreichen Revisionsprozesses. 1st ed. Berlin: Erich Schmidt Verlag (DIIR-Forum, 10).

COSO – The Committee of Sponsoring Organizations of the Treadway Commission (2004): Unternehmensweites Risikomanagement – Übergreifendes Rahmenwerk.

DIIR – Deutsches Institut für Interne Revision e. V (2000): Revision von Architekten- und Ingenieurleistungen (HOAI). 2., überarb. und erw. Aufl. Berlin: Erich Schmidt Verlag (IIR-Schriftenreihe/Deutsches Institut für Interne Revision e.V., Frankfurt, M, 19).

DIIR – Deutsches Institut für Interne Revision e. V (2006): Revision des Facility-Managements. Ein Prüfungsleitfaden. Berlin: Erich Schmidt Verlag (IIR-Schriftenreihe/Deutsches Institut für Interne Revision e.V., Frankfurt, M, 38).

DIIR – Deutsches Institut für Interne Revision e. V (2010): Revision von Bauleistungen. Kommentierte Prüfungsfragen für die Revisionspraxis. 3., neu bearb. Aufl. Berlin: Erich Schmidt Verlag (IIR-Schriftenreihe/Deutsches Institut für Interne Revision e.V., Frankfurt, M, 6).

DIIR – Deutsches Institut für Interne Revision e. V (2013): Revision der Instandhaltung. Bauwerke – Technische Anlagen – Außenanlagen. Berlin: Erich Schmidt Verlag (DIIR-Forum, 48). Online verfügbar unter https://site.ebrary.com/lib/alltitles/docDetail.action?docID=10697046.

DIIR – Deutsches Institut für Interne Revision e. V (2015): Schutz vor dolosen Handlungen bei Bauprojekten. Leitfaden für eine risikoorientierte Prozessanalyse. 1st ed. Berlin: Erich Schmidt Verlag (DIIR-Schriftenreihe, 56).

DIIR – Deutsches Institut für Interne Revision e. V (2017): Revision des Claimmanagements. Leitfaden zur prozessorientierten Prüfung von Nachträgen bei Bauprojekten. Berlin: Erich Schmidt Verlag (DIIR-Schriftenreihe, v.58).

DIIR – Deutsches Institut für Interne Revision e. V (2020): Aufgaben und Ziele des Arbeitskreises „Bau, Betrieb und Instandhaltung". DIIR – Deutsches Institut für Interne Revision e. V. Online verfügbar unter https://www.diir.de/arbeitskreise/bau-betrieb-und-instandhaltung/aufgaben-und-ziele/, zuletzt geprüft am 09.10.2020.

P. Wotschke, G. Kindermann, *Construction Auditing*,
https://doi.org/10.1007/978-3-658-38841-6

DIIR Arbeitskreis MaRisk (Hg.) (2019): Online-Revisionshandbuch für die Interne Revision in Kreditinstituten. DIIR – Deutsches Institut für Interne Revision e. V.

Dudenredaktion (2020): „Qualität" auf Duden online. Online verfügbar unter https://www.duden.de/node/116878/revision/243450, zuletzt aktualisiert am 01.10.2020, zuletzt geprüft am 01.10.2020.

Düsterwald, Robert (2010): Leitfaden zur Prüfung von Projekten. Erläuterungen und Empfehlungen zum DIIR Standard Nr. 4. Berlin: Erich Schmidt Verlag (DIIR-Schriftenreihe, Bd. 45). Online verfügbar unter https://site.ebrary.com/lib/alltitles/docDetail.action?docID=10628633.

Eller, Roland (2010): Kompaktwissen Risikomanagement. Nachschlagen, verstehen und erfolgreich umsetzen. 1. Aufl. Wiesbaden: Gabler Verlag/Springer Fachmedien Wiesbaden GmbH Wiesbaden.

Eulerich, Marc (2018): Die interne Revision. Theorie – Organisation – Best Practice. Berlin: Erich Schmidt Verlag.

Gleißner, Werner (2016): Reifegradmodelle und Entwicklungsstufen des Risikomanagements: ein Selbsttest. In: Controller Magazin (11), S. 31–36.

Hirzel, Matthias; Gaida, Ingo; Geiser, Ulrich (Hg.) (2013): Prozessmanagement in der Praxis. Wertschöpfungsketten planen, optimieren und erfolgreich steuern. 3., überarb. u. erw. Aufl. 2013. Wiesbaden: Gabler Verlag.

Hoffmann, Wilfried (2017): Risikomanagement. Kurzanleitung Heft 4. 2nd ed. Berlin, Heidelberg: Springer Vieweg (DVP Projektmanagement).

Hunziker, Stefan; Meissner, Jens O. (Hg.) (2018): Ganzheitliches Chancen- und Risikomanagement. Interdisziplinäre und praxisnahe Konzepte. Wiesbaden: Springer Gabler.

ISACA Germany Chapter (2014): Leitfaden ISO 31000 in der IT.

Jakoby, Walter (2019): Qualitätsmanagement für Ingenieure. Ein praxisnahes Lehrbuch für die Planung und Steuerung von Qualitätsprozessen (Lehrbuch).

Kübel, Moritz (2013): Corporate M&A. Reifegradmodell und empirische Untersuchung. Zugl.: Erlangen-Nürnberg, Univ., Diss., 2011. Wiesbaden: Springer Gabler (Springer Gabler Research).

Madauss, Bernd-J. (2017): Projektmanagement. Theorie und Praxis Aus Einer Hand. 7th ed. Berlin, Heidelberg: Springer Vieweg.

Matschechin, Alexander (2017): Vergleich verschiedener Möglichkeiten zur Prozessanalyse und Prozessdarstellung. Hochschule für Angewandte Wissenschaften Hamburg, Hamburg. Department Wirtschaft.

Oepen, Ralf-Peter (Hg.) (2012): Risikoorientierte Bauprojekt-Kalkulation. Eine innovative Methode zur Risikobeherrschung und Eindämmung von Ausreißer-Projekten. BRZ Deutschland GmbH. 1. Aufl. Wiesbaden: Vieweg & Teubner.

Otremba, Stefan (2016): GRC-Management als interdisziplinäre Corporate Governance. Dissertation.

PICTURE GmbH (2020): Prozessmanagement mit der PICTURE-Methode. Analyse und Optimierung von Geschäftsprozessen als Mittel der Organisationsgestaltung. Münster. 91.

DIN EN ISO 90001, 2015: Qualitätsmanagementsysteme.

DIN ISO 31000, 2011: Risikomanagement.

Risner, Ron (2012a): 1105dl_Practitioners-Blueprint-to-Construction-Auditing_Deutsche Übersetzung.

Risner, Ron (2012b): The practitioner's blueprint to construction auditing. Altamonte Springs, Fla.: Inst. of Internal Auditors Research Foundation.

Röglinger, Maximilian; Kamprath, Nora (2012): Prozessverbesserung mit Reifegradmodellen. Eine Analyse ökonomischer Zusammenhänge. In: Zeitschrift für Betriebswirtschaft (5).

Romeike, Frank (2018): Risikomanagement. Wiesbaden: Springer Gabler (Studienwissen kompakt).

Schleinzer, Andreas (2014): Entwicklung eines Reifegradmodells für das unternehmensweite Risikomanagement. Diplomarbeit. Technische Universität, Wien. Institut für Managementwissenschaften.

Schmidt, Christoph (2016): Steigerung der Objektivität Interner Revisoren. Dissertation.

Schmidt, Günter (2012): Prozessmanagement. Modelle und Methoden. 3. überarb. Aufl. 2012. Berlin: Springer Gabler.

Schneider, Werner (2017): Prozessorientiertes Bauprojektmanagement. Kurzanleitung Heft 1. 3., neu bearbeitete Auflage. Hg. v. Walter Volkmann. Berlin, Heidelberg: Springer Vieweg (DVP Projektmanagement).

Schwager, Elmar; Fischer, Josef (2008): Prüfungs- und Beratungsleistungen einer modernen Baurevision. Ein Überblick zu aktuellen Prüffeldern, Beratungsleistungen und der Bekämpfung von Fraud 2008.

Seidlmeier, Heinrich (2019): Prozessmodellierung mit ARIS. Eine beispielorientierte Einführung für Studium und Praxis in ARIS 10: Springer Vieweg.

Transparency International e.V. (Hg.) (2005): Jahrbuch Korruption 2005. Schwerpunkt: Bau und Wiederaufbau. Korruption im Bausektor: ein Blick auf das Fundament. Unter Mitarbeit von Neill Stansbury: Partlas Verlag.

Westhausen, Hans-Ulrich (2016): Interne Revision in Verbundgruppen und Franchise-Systemen. Dissertation.

Wingsch, Dittmar (2005): Baurevision in der Praxis. In: Interne Revision (2), S. 64–66.

Printed in the United States
by Baker & Taylor Publisher Services